Sustainable Highways, Pavements and Materials

An Introduction

Kasthurirangan Gopalakrishnan, Ph.D.

Transdependenz LLC

Dedicated to my
Teachers
with utmost gratitude

Contents

Preface

Sustainability is a key issue facing today's society and arguably the hottest topic of the moment. There is increasing awareness that certain human activities and development practices are negatively affecting the planet's well-being and threatening the existence of the future generations. Most notably, the burning of fossil fuels for energy production and transport are implicated in forcing global climate change and increased Greenhouse Gas (GHG) emissions with extreme environmental and human health impacts. Some of the more prominently cited examples of such impacts include rising sea levels, shrinking glaciers, water scarcity, changes in the range and distribution of plants and animals, and seasonal irregularities.

The interconnectedness of the social, economic and environmental linkages is increasingly becoming evident. In essence, sustainability strives to balance the three E's, namely Environmental protection, Economic vitality, and Equal opportunity (equity), in meeting the needs of today without compromising those of tomorrow. The need for a sustainable future world has brought together the policymakers, environmentalists, the business leaders, scientists, educators, and the people in general under a common umbrella to formulate guidelines, policies and initiatives promoting green principles and minimizing negative environmental impacts.

Highway agencies and engineers are also rising to the occasion by actively promoting sustainability-oriented engineering practices and projects. Terms like green highways, sustainable pavements, carbon footprint, environmental Life-Cycle Assessment (LCA) etc. are beginning to make increased appearances in transportation literature, and in the titles of conferences and peer-reviewed journal special issues.

Sustainability is not an entirely new concept in road infrastructure. While economical considerations and project-specific constraints dominated the scene previously, the current push towards environmental sustainability, in addition to financial pressures, is mainly coming from impending governmental regulations and societal pressures on highway authorities to preserve the environment and ecosystems. Almost every transportation agency's website now has a prominent section describing the sustainability initiatives currently being undertaken by the agency towards implementing sustainable transportation practices and declaring its commitment to environmental stewardship.

Although newer and newer technologies are being introduced in the design, construction, and maintenance of highways and pavements, there are no standard guidelines on incorporating sustainability concepts in to any of these processes nor are there best-practices guidelines on sustainable practices. At the same time, many educators and organizations have already stressed the paramount importance of formally incorporating sustainability concepts in the civil engineering curriculum to educate and train a new class of civil engineers well-equipped to address the future sustainability challenges. This book is an attempt to partially fill these needs by synthesizing the existing relative information on sustainable highway infrastructure in one place to serve as a useful textbook and reference to researchers, practitioners, educators, as well as students with an interest in transportation infrastructure systems. Like the topic itself which the book addresses, the content of the book is also expected to evolve with time although the overall framework has been established in this first edition to provide an adequate introduction.

Chapters 1 and 2 lay the foundation by briefly introducing the origins of sustainability and sustainable development and some of the useful sustainability definitions. These definitions underline the need for living within the limits and call for a better understanding of the interdependency among ecological, economic and social systems. "Sustainable highways and pavements" is introduced as a distinct and yet an integral component of "sustainable transportation". Many of the inherent characteristics of pavements and existing pavement practices can be used to enhance the sustainability of both asphalt and concrete pavements, the two prominent pavement types. Such characteristics include the opportunities to incorporate recycled materials and industrial residuals and build long-life pavements that are safe, quiet, durable and economical with surfaces that minimize urban heat and harvest and purify rainwater.

Chapter 3 highlights some of the prominent sustainable transportation initiatives which address not only the environmental and natural resource needs, but also focus on building roadways that are safer, more livable and accessible and welcoming to everyone, not just the automobile traffic.

Chapter 4 begins by presenting some well-known alarming facts about the various negative impacts highways and pavements have on the environmental, economic, and social dimensions of the society when viewed from a life-cycle perspective. The environmental LCA framework is introduced and the different LCA approaches are briefly discussed. Finally, the existing pavement-related LCA tools and GHG calculators are briefly summarized.

A crucial understanding emerging in the sustainable pavements arena is that there are several points in the lifecycle of a pavement where sustainability concepts can be potentially applied. From design to construction to operation to maintenance and rehabilitation and end-of-life of pavements, there is opportunity at every stage to preserve or enhance economic, environmental and social well-being by more actively adopting strategies like recycling and beneficial re-use of industrial by-products which have been employed in highway practice for decades. Many of these practices and technologies are receiving renewed attention from the highway community with the current emphasis on environmental sustainability.

Chapters 5 to 10 introduce some of the more commonly known sustainability aspects of pavements such as the use of recycled materials, long-life pavements, Warm-Mix Asphalt (WMA), and pavements that often have the potential to simultaneously address the problems related to stormwater runoff, highway traffic and construction noise, and heat island. These are referred to as permeable, quiet, and cool pavements.

Chapter 11 provides an up-to-date discussion on the existing and emerging roadway sustainability assessment and rating/certification systems, which are often referred to as "green" rating systems. These systems are intended to measure the sustainability of highway projects and they work by recognizing and awarding more points to sustainability strategies that will have greater positive impacts on energy efficiency and CO_2 emissions. Finally, Chapter 12 concludes with an overview of the emerging sustainable pavement materials and technologies.

This edition is also brought out in the form of an e-book to facilitate instant access to the material as well as to allow readers to use hyperlinks to more easily take advantage of the plethora of online resources that appear throughout the book. Apart from the useful online resources listed at the end of every chapter, almost every reference included in the book has an associated web link through which one may be able to access or download the referenced material.

The readers are requested to take note of the fact that this book was originally published in 2011 and much has changed since then and will continue to change, especially with respect to the development of sustainable highway and pavement rating systems.

Acknowledgments

This book has heavily drawn on the work of others and it has not always been possible to acknowledge this in detail to avoid presentation difficulties. The aim was not to create new knowledge as such, but present available information in collected, collated and readable form. It is sincerely apologized if this manner of presentation has led to any perceived inadequacies in dereferencing of the work of others due to unavoidable reasons.

First and foremost, the author wishes to gratefully acknowledge the following organizations for making significant resources on sustainability related topics freely available online and for generously giving their permission to use a number of figures included in this book: US Environmental Protection Agency (EPA), Federal Highway Administration (FHWA), State Departments of Transportation (DOTs), Intergovernmental Panel on Climate Change (IPCC), US National Oceanic and Atmospheric Administration's (NOAA), National Aeronautics and Space Administration (NASA), US National Park Service (NPS), American Society of Civil Engineers (ASCE), Transportation Research Board (TRB), American Association of State Highway and Transportation Officials (AASHTO), Green Highways Partnership (GHP), U.K. Roads Board, RILEM, City of Charlotte, National Complete Streets Coalition, US

Geological Survey (USGS), Lawrence Berkeley National Laboratory (LBNL) and International Organization for Standardization (ISO).

The documents, research reports and illustrations provided by the following organizations were extremely useful in preparing this book: American Concrete Pavement Association (ACPA), Asphalt Pavement Alliance (APA), National Asphalt Pavement Association (NAPA), European Asphalt Pavement Association (EAPA), Portland Cement Association (PCA), National Center for Asphalt Technology (NCAT), National Concrete Pavement Technology Center (CP Tech Center), United States Green Building Council (USGBC), Victoria Transport Policy Institute (VPTI), Ontario Ministry of Transportation (MTO), Industrial Resources Council (IRC), and Recycled Materials Resource Center (RMRC). Due to space, a number of other organizations under this category could not be individually acknowledged.

The author is grateful to a large number of people, including his advisors, mentors, friends, colleagues, co-authors, and students who have helped him to develop his thinking about sustainability and pavements over the years and for their contributions towards this book directly or indirectly.

Finally, the author cannot fail to acknowledge with deep gratitude the contributions made by every individual whose work is referenced in this book.

Chapter 1:
Introduction

A Brief History of Sustainability

The concept of sustainability is as old as some of the oldest cultures on Earth which aimed for stability and perpetuity (1). In those cultures, qualities like humility and contentment were encouraged and respect for tradition was stressed by elders who were more concerned about the long-term consequences of their action, than short-term gains (2,3). For instance, the Vedas (body of knowledge originating in ancient India) hold that "upon a correct evaluation of the needs of the self, an individual will be completely satisfied and will therefore not desire to live and consume in a way that is destructive to their surrounding environment and its inhabitants" (4). Similarly, the Ancient Greeks recognized that "a society grows great when old men plant trees in whose shade they know they shall not sit" (5). The Constitution of the Iroquois Nations (of the Native Americans) states, "look and listen for the welfare of the whole people and have always in view not only the present but also the coming generations, even those whose faces are yet beneath the surface of the ground – the unborn of the future Nation" (6).

Although change still happened with time, it was not sought and "people saw that innovation was more likely to bring peril than progress, and even more curiosity could open a Pandora's box of trouble" (1). Over the last few hundred years, this concern was gradually abandoned in favor of progress through scientific, technological and economic advances accompanied by major unintended changes in the ecosystem (1,7).

The resurgence of the old sustainability idea in a new form in the latter part of the 20th century has been spurred by a series of incidents and discoveries, including the meteorological problems caused by El Nino, the melting of the glaciers, climate change and global warming, the hole in the Antarctic ozone layer (see Fig. 1.1), the disappearance of many animal and plant species, contamination of the food chain with mercury and other metals, the increasing salinity of soil in many agricultural areas, escalating rates of soil erosion, loss of permeability of soil, the shrinkage of large bodies of water such as the Aral sea, rising ocean acidity, etc (8). Since then, a general awareness of human-induced climate change and greenhouse gases and their potential environmental and socio-economic impacts has increased.

The modern contemporary view of sustainability draws its roots from six separate strains of thought that have emerged prominently since the 1950 namely the ecological/carrying-capacity root, the critique of technology root, the biosphere root, the resources/environment root, the no-growth/slow-growth root, and the eco-development root. These roots are embedded in fundamentally different concepts and values and have arisen from discussions on interrelationships among population growth rates, resource use, and pressure on the environment (10). This has led to diverse viewpoints regarding the meaning of sustainability although this uncertainty has not reduced the popularity of the concept (11).

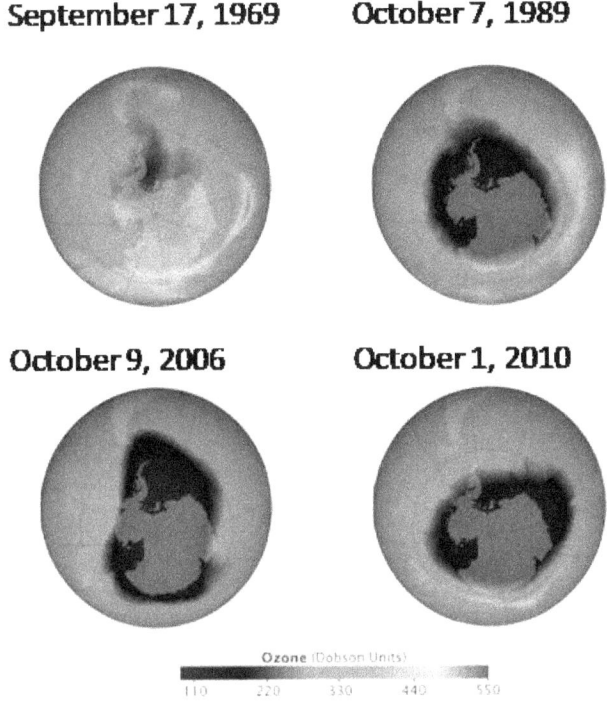

Fig. 1.1. Antartic ozone hole through the years on the day of its maximum depletion in four different years [*Source:* NASA (9)]

In the debate about sustainable development, there is a frequent reference to two mutually exclusive visions and opposing paradigms of sustainability depending on the costs incurred in attaining them: strong sustainability and weak sustainability (12).

Strong sustainability equates to ecological sustainability where the focus is primarily on the environment without accepting any kind of trade-off between economic gain and environmental quality. The proponents of strong sustainability maintain that man-made capital and natural capital (i.e., natural resources that provide us with ecosystem goods and services) are not substitutable since any damage to the environment will have negative repercussions (11).

Weak sustainability equates to a sort of economic sustainability supported by neo-classically oriented economists who argue that different forms of capital (natural, man-made, social, financial etc) are substitutable with one another and therefore environmental quality can be traded against economic gain (11). In other words, weak sustainability "allows for the depletion or degradation of natural resources, so long as such depletion is offset by increases in the stocks of other forms of capital (for example, by investing royalties from depleting mineral reserves in factories)" (13). Of the two, the weak sustainability is the one that presently dominates in the global economy (11).

Irrespective of the differing notions and perspectives related to sustainability, there is a general agreement that humans are implicated in forcing climate change and there is a scientific consensus that all nations must come together to reduce the greenhouse gas emissions and consequently the planet's energy footprint.

Greenhouse Gases (GHGs) and Climate Change

Greenhouse Gases (GHGs) are gases that trap heat in the atmosphere. When sunlight strikes the Earth's surface, most of the energy radiated by the Sun is absorbed by the Earth, while some solar radiation is reflected by the Earth back towards space as infrared radiation. The GHG molecules in the atmosphere trap the heat by absorbing much of this energy going back to space and radiate most of it back to the Earth's surface creating a natural "green house effect" (see Fig. 1.2). These GHGs greatly affect Earth's temperature and it is believed that without them, Earth's surface would be about 60 °F lower than they are now.

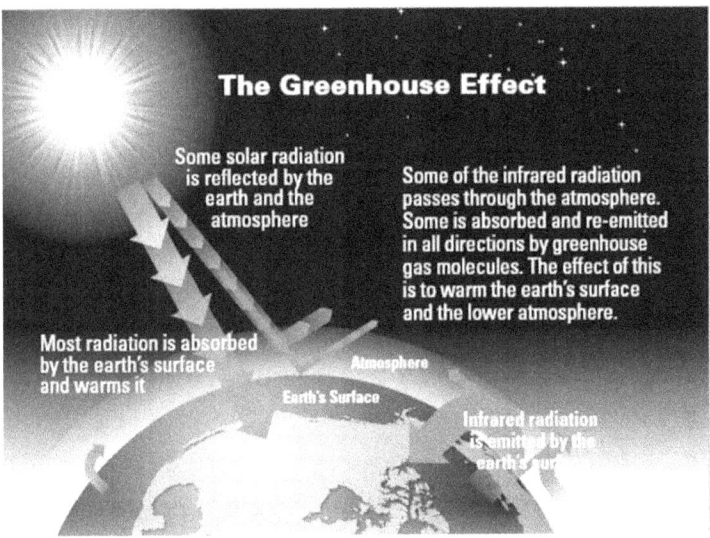

Fig. 1.2. The natural greenhouse effect [*Source:* US EPA (14)]

Over time, the temperature of the Earth's surface would be roughly constant if atmospheric concentrations of GHGs remain relatively stable allowing the amount of energy sent from the Sun to the Earth's surface to be about the same as that radiated back into space (14). But, this is not to be so. According to the National Oceanic and Atmospheric Administration's (NOAA) State of the Climate Report and the National Aeronautics and Space Administration's (NASA) Surface Temperature Analysis, the indications are that the average temperature of the Earth's surface has increased by about 1.2 to 1.4 deg-F since 1900 (14).

The Intergovernmental Panel on Climate Change (IPCC), formed jointly in 1988 by the United Nations Environment Programme and the United Nations World Meteorological Organization, produces authoritative assessments of the current state of knowledge of climate change by bringing together world's top scientists, economists, and other experts. The *2007 Fourth Assessment Report* compiled by the IPCC confirms that "changes in atmospheric concentrations of greenhouse gases and aerosols, land cover and solar radiation alter the energy balance of the climate system". It was also noted that many natural systems are being affected all over the world by temperature increases.

The IPCC's *Fourth Assessment Report* concluded that "increases in anthropogenic greenhouse gas concentrations is very likely to have caused most of the increases in global average temperatures since the mid-20th century". Anthropogenic refers to that originating from human activities. In other words, most of the global warming in recent decades is very likely the result of activities of humans (see Fig. 1.3).

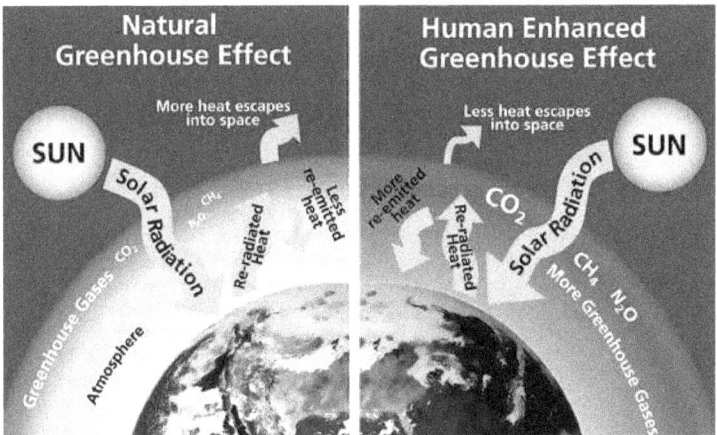

Fig. 1.3. Illustration of natural greenhouse effect and increasing greenhouse gas levels resulting from human activities [*Source:* W. Elder, NPS (15)]

The primary and natural GHGs in the Earth's atmosphere include water vapor (H_2O), carbon dioxide (CO_2), methane (CH_4), nitrous oxide (N_2O), and ozone (O_3). There are other industrial GHGs which are exclusively man-made. Apart from CO_2, CH_4, and N_2O, fluorinated gases (also referred to as F-gases or ozone-depleting substances [i.e., CFCs, HCFCs, and halons]), which include hydrofluorocarbons, perfluorocarbons, and sulfur hexafluoride, are also included in the list of principal anthropogenic GHGs. These GHGs enter the atmosphere, mainly through the burning of fossil fuels (such as coal, natural gas, oil and gasoline for industrial and transport applications) and clearing of forests. Fig. 1.4 shows global anthropogenic GHG emissions from 1970 to 2004.

"Carbon footprint" and "ecological footprint" are two terms that are used more often to measure the impact of our activities on the environment and climate change. Carbon footprint describes the total amount of GHGs that are emitted into the atmosphere each year caused directly or indirectly by an entity such as a person, household, building, organization, or company (16). It is measured in tons of CO_2 equivalent allowing different GHGs to be compared relative to one unit of CO_2. The ecological footprint is a measure of human demand on the Earth's ecosystems and natural resources. As an indicator of environmental sustainability, it compares the natural resources consumed by a human population with Earth's ecological capacity to regenerate them (17).

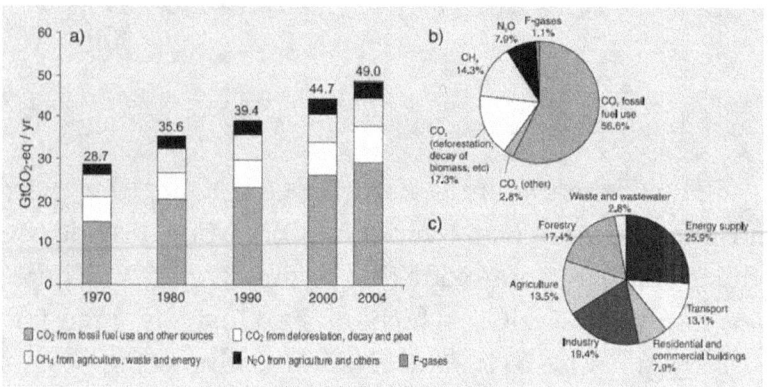

Fig. 1.4. (a) Global annual emissions of anthropogenic GHGs from 1970 to 2004; (b) Share of different anthropogenic GHGs in total emissions in 2004 in terms of CO_2-equivalent; (c) Share of different sectors in total anthropogenic GHG emissions in 2004 in terms of CO_2-eq. [Source: IPCC Climate Change 2007: Synthesis Report, Fig. 2.1, p. 36 (18)]

Carbon dioxide (CO_2) is considered the most prevalent of all GHGs and other GHG emissions are typically reported in terms of CO_2 equivalent (CO_2 Eq.) to provide a common unit of measure. CO_2 is produced by both natural and human activities, and like most GHGs, can be removed (or sequestered) through natural processes. However, the total GHGs have exceeded the natural absorption rates and their atmospheric concentrations have increased because of increased production of CO_2 by human sources (18,19).

To either mitigate or defer global warming, industrially produced CO_2 is also artificially sequestered by capturing it using subsurface reservoirs, aquifers, or other carbon sinks. Carbon Capture and Storage (CCS), also referred to as carbon capture and sequestration is a mitigation measure for reducing the fossil fuels emissions contribution to global warming by capturing CO_2 from the sources and storing it without allowing it to enter the atmosphere (20). Many believe that climate change mitigation (i.e., emission reduction, renewable energy, etc.) and adaptation and a low-carbon economy are critical for a sustainable future world development.

Sustainable Development

In 1987, the World Commission on Environment and Development (WECD) published a report titled *Our Common Future* (often called the *Report of the Brundtland Commission* of the United Nations [UN]) which highlighted the need for economic and social development without depleting natural resources or harming the environment (7). It defined sustainable development as "development that meets the needs of the present without compromising the ability of future generations to meet their own needs" (7).

Although several re-definitions and interpretations of sustainable development have evolved since then, the Brundtland report's (7) definition of sustainable development has been widely accepted by most policy-makers and institutions across the world. The three fundamental components to sustainable development highlighted by this report include: environmental protection, economic growth, and social equity. These three issues have come to be known as the three inseparable Es of sustainable development which also form the "sustainability triangle" (see Fig. 1.5).

The "triple bottom line" (TBL or 3BL) (also known as "the three pillars") of sustainability is succinctly described by these three categories: social equity (people), environment (planet), and economy (profit). The idea is to not to undervalue any one category, but work to achieve long-term well-being by balancing community, environment, and economy. Sustainable development is not intended to reflect the trade-off between economy and the environment, but the synergy between them facilitated by public participation (21). Sustainability is increasingly being looked at with a broader lens to include a range of often overlapping issues identified in Fig. 1.6.

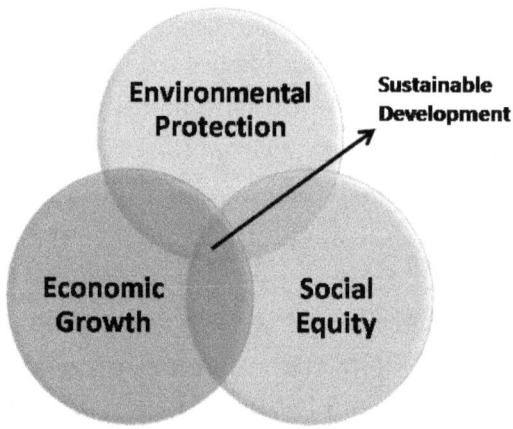

Fig. 1.5. The three inseparable "E"s of sustainable development

In December 1997, after a decade of international meetings and negotiations, more than 160 countries installed the Kyoto Protocol or Kyoto Climate Agreement (following the UN Framework Convention on Climate Change), committing themselves to reduce heat trapping GHG emissions and address the problem of climate change. According to the Protocol, 38 industrialized nations agreed to reduce their total GHG emissions by roughly five percent by 2012 compared with the 1990 levels (23). The Kyoto Protocol went into effect on February 16, 2005. A total of 178 countries and other governmental entities have ratified the agreement as of April 2008.

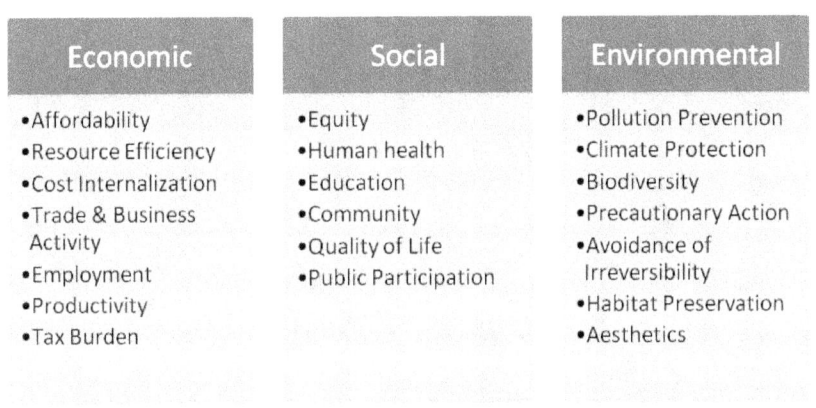

Fig. 1.6. Overlapping sustainability issues [*Source:* VPTI (22)]

It is estimated that the atmospheric concentrations of CO_2 have increased by nearly 30 percent since the beginning of the industrial revolution. Anthropogenic atmospheric CO_2, fossil fuel emissions, world gross domestic product (GDP), and world population from 1900-2008 are plotted together in Fig. 1.7. It is seen that atmospheric CO_2 continues to grow despite the recent global economic slump that led to decrease in activities related to the use of fossil fuels (24).

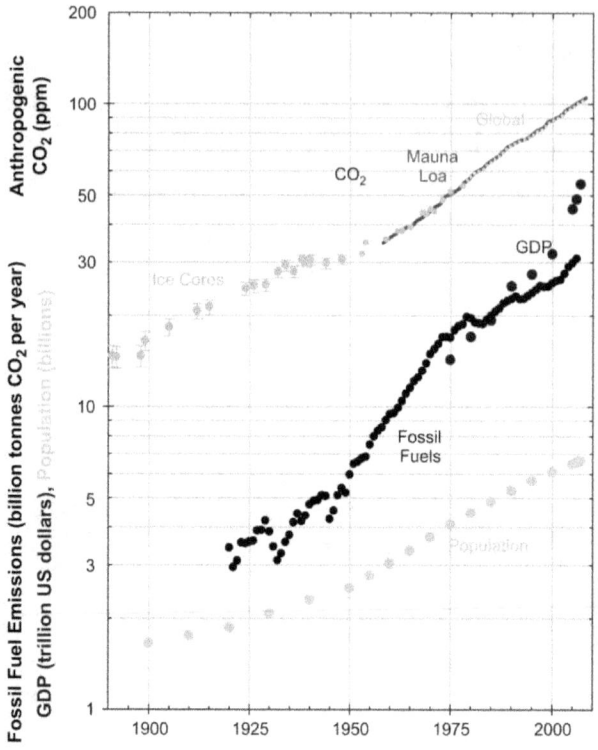

Fig. 1.7. Anthropogenic atmospheric CO_2, fossil fuel emissions, world gross domestic product (GDP), and world population for the past century [*Source:* NOAA (24)]

The combustion of fossil fuels is considered a major contributor to anthropogenic (i.e., resulting from human activities) CO_2 emissions, with land-use change providing a significant, but smaller contribution. The transportation sector which derives most of its energy from fossil fuels is estimated to be responsible for over 20% of the global GHG emissions, of which some 85% is attributed to activities resulting from road transportation alone. Similarly, road transportation accounts for 72% of all transportation related

CO_2 emissions in the European Union (EU) member countries (25). Among the world's transportation sectors, the U.S. transportation sector is by far the largest source of GHG emissions.

Sustainable Highways and Pavements

Highways are the backbone of any nation's transportation system and they form the foundation of a nation's economic prosperity providing the ability to safely and efficiently transport people, products, and goods (26). The impact of highways on the environment is huge resulting from its life cycle activities which involve planning, design, material extraction, construction, usage, maintenance, and demolition (27).

A highway transportation project can affect the natural ecosystems through alteration of topography, removal of vegetation, damage of habitat and bio-diversity, erosion, sedimentation, and soil compaction, stormwater pollution, increased GHG emissions and warming, noise and visual disturbance, etc. to name a few (27,28). Fig. 1.8 illustrates the interrelationships among social and economic effects of a proposed transportation project.

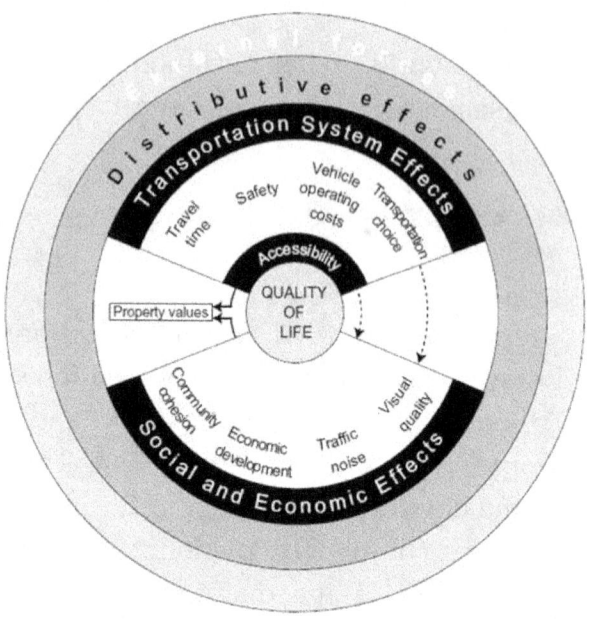

Fig. 1.8. Interrelationships among social and economic effects of a proposed transportation project [*Source:* NCHRP (29)]

Acknowledging that transportation projects can have substantial effects which can alter social, economic, and cultural systems and processes, the US National Environmental Policy Act (NEPA) of 1969, for instance, mandated a formal process for evaluating potential impacts to social and natural environment when considering approval of proposed transportation projects. The US Department of Transportation (DOT) and the Federal Highway Administration (FHWA) hold similar positions in their strategic goal for the environment (30). The FHWA NEPA project development process was initiated to allow transportation officials "to make project decisions that balance engineering and transportation needs with social, economic, and natural environmental factors". A wide range of partners including the public, businesses, interest groups, and governmental agencies provide input into project and environmental decisions during this decision making process (31).

The road pavement is a fundamental element integral to highway infrastructure. More often, the line between a pavement and a highway/roadway is indiscernible, although a distinction is made in this book with the recognition that a road pavement is a subset of highway/roadway infrastructure. Pavements are typically constructed in distinct layers composed of materials such as asphalt, cement, aggregates, and soil. Fig. 1.9 shows the life cycle phases of a typical pavement system.

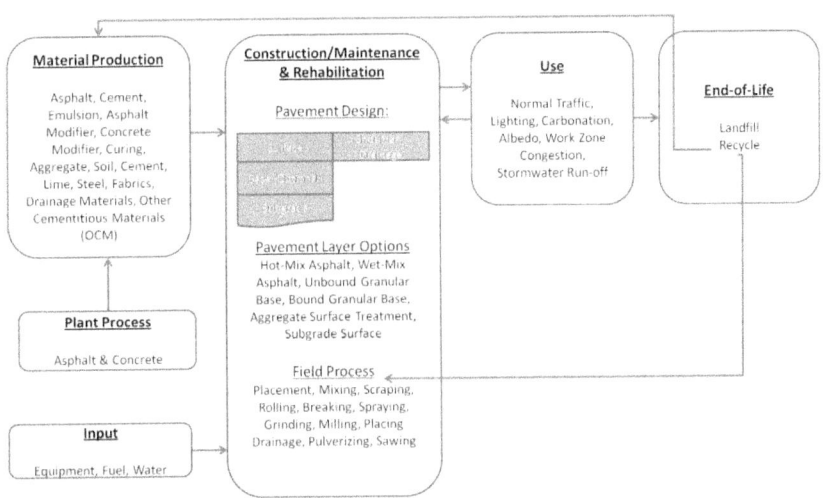

Fig. 1.9. Life cycle phases of a typical pavement system [Source: UCPRC P-LCA Team (32)]

Although the principles of sustainability are widely implemented in other building and construction industries with standard provisions, these principles are not yet widely understood or implemented in highway pavement design and construction. This book is an attempt to partially fill this gap by first providing an overall introduction to this topic for the benefit of students, researchers and practitioners.

Useful Online Resources

Climate Change and Greenhouse Gas Emissions

The Intergovernmental Panel on Climate Change (IPCC) is the leading international body for the assessment of climate change

http://www.ipcc.ch/

US Environmental Protection Agency's (EPA's) Climate Change Programs and Activities

http://www.epa.gov/climatechange/index.html

The US National Oceanic and Atmospheric Administration (NOAA) is a federal agency focused on the condition of the oceans and the atmosphere

http://www.noaa.gov/

The World Meteorological Organization (WMO) is the United Nation (UN) system's authoritative voice on the state and behavior of the Earth's atmosphere and its interaction with the oceans http://www.wmo.int/pages/about/index_en.html

National Aeronautics and Space Administration's (NASA's) Eyes on the Earth: Global Climate Change

http://climate.nasa.gov/

Greenhouse Gas Related Web Links

http://www.ghgonline.org/weblink.htm

Sustainability

The Sustainability Program of US Environmental Protection Agency's (EPA's) Office of Research and Development

http://www.epa.gov/sustainability/

Report of the World Commission on Environment and Development: *Our Common Future* (Report of the Brundtland Commission)

http://www.un-documents.net/wced-ocf.htm

UN Department of Economic and Social Affairs: Division for Sustainable Development

http://www.un.org/esa/dsd/index.shtml

A list of internet websites related to sustainable development

http://www.ulb.ac.be/ceese/meta/sustvl.html

Highways and Pavements

The US Department of Transportation (DOT) oversees federal highway, air, railroad, and maritime and other transportation administration functions

http://www.dot.gov/

The Federal Highway Administration (FHWA) is an agency within the US DOT that supports State and local governments in the design, construction, and maintenance of the US highway system (Federal Aid Highway Program) and various federally and tribal owned lands (Federal Lands Highway Program)

http://www.fhwa.dot.gov/

http://www.fhwa.dot.gov/pavement/

The UK Highways Agency is an Executive Agency of the Department for Transport (DfT), and is responsible for operating, maintaining and improving the strategic road network in England on behalf of the Secretary of State for Transport

http://www.highways.gov.uk/default.aspx

Transport Canada (TC) is responsible for transportation policies and programs in Canada and ensures that air, marine, road and rail transportation are safe, secure, efficient and environmentally responsible http://www.tc.gc.ca/eng/tc-main.htm

Pavement Interactive: The online community for all things related to pavements

http://pavementinteractive.org/

The Long-Term Pavement Performance Program (LTPP) began in 1987 with a series of rigorous long-term field experiments monitoring more than 2,400 asphalt and portland cement concrete pavement test sections across the U.S. and Canada

http://www.fhwa.dot.gov/research/tfhrc/programs/infrastru cture/pavements/ltpp/index.cfm

The Mechanistic-Empirical Pavement Design Guide (MEPDG), related documentation, and the latest version of its software are available online for evaluation through 30 September 2011

http://onlinepubs.trb.org/onlinepubs/archive/mepdg/home.h tm

References

1. Gibson RB, Hassan S. Sustainability assessment: criteria and processes. Earthscan; 2005. 268 p.

2. Ab Razak RR, Sanusi ZA. The Concept of Sustainable Development in Human Civilisation: An Introspective View. KEMANUSIAAN: The Asian Journal of Humanities [Internet]. 2010 [cited 2011 Feb 28];17:19-36.

Available from:
http://www.usm.my/kajh/KAJH%2017/KAJH%20Art%2
02%20(19-36).pdf

3. Burdyuzha V, Kohzin G, editors. The Future of the
    Universe and the Future of Our Civilization. Singapore:
    World Scientific Publishing Co. Pte. Ltd.; 2000. 408 p.

4. Stewart-Jacks SD. Sustainable Accountability:A Dualist
    Vedic Perspective [Internet] [M.S. Thesis]. [New
    Zealand]: Victoria University of Wellington; 2009 [cited
    2011 Feb 28]. 149 p.Available from:
    http://researcharchive.vuw.ac.nz/bitstream/handle/1006
    3/933/thesis.pdf?sequence=1

5. McVoy GR, Nelson DA, Krekeler P, Kolb E, Gritsavage
    JS. Moving Towards Sustainability:New York State
    Department of Transportation's GreenLITES Story
    [Internet]. In: Proceedings of the Green Streets and
    Highways 2010 Conference. Denver, Colorado: ASCE
    Press; 2010 [cited 2011 Feb 28]. p. 20.Available from:
    https://www.nysdot.gov/programs/greenlites/repositor
    y/ASCE%20GreenLITES%20Final%20Paper%207-12-
    10.pdf

6. Murphy G, Welker G. The Constitution of the Iroquois
    Nations: The Great Binding Law [Internet]. 2007 [cited
    2011 Feb 28];Available from:
    http://www.indigenouspeople.net/iroqcon.htm

7. Brundtland GH. Our Common Future: Report of the
    World Commission on Enviornment and Development
    [Internet]. Oslo, Norway: United Nations (UN); 1987
    [cited 2011 Feb 27]. Available from: http://www.un-
    documents.net/wced-ocf.htm

8. US EPA. Climate Change Indicators in the United States [Internet]. [cited 2011 Feb 27];Available from: http://www.epa.gov/climatechange/indicators/pdfs/climate_indicators_slideshow.pdf

9. NASA. Ozone Hole through the years : Image of the Day [Internet]. NASA Earth Observatory. 2011 Jan 31 [cited 2011 Feb 28];Available from: http://earthobservatory.nasa.gov/IOTD/view.php?id=49040

10. Kidd CV. The evolution of sustainability. J Agric Environ Ethics. 1992;5(1):1-26.

11. Bell S, Morse S. Sustainability indicators: measuring the immeasurable? Second Edition. London, UK: Earthscan; 2008. 251 p.

12. Neumayer E. Weak versus strong sustainability: exploring the limits of two opposing paradigms. Second Edition. United Kingdom: Edward Elgar Publishing Ltd.; 2003. 294 p.

13. OECD. OECD Glossary of Statistical Terms - Weak sustainability Definition [Internet]. 2005 Jul 6 [cited 2011 Mar 23];Available from: http://stats.oecd.org/glossary/detail.asp?ID=6611

14. US EPA. Frequently Asked Questions About Global Warming and Climate Change: Back to Basics [Internet]. Washington, D.C.: US Environmental Protection Agency (EPA); 2009 [cited 2011 Feb 27]. Available from: http://epa.gov/climatechange/downloads/Climate_Basics.pdf

15. National Park Service (NPS). What is Climate Change? [Internet]. 2010 May 22 [cited 2011 Feb 28];Available

from: http://www.nps.gov/goga/naturescience/climate-change-causes.htm

16. US EPA. Climate Change: Frequent Questions - What is a carbon footprint? Where did this term originate? [Internet]. US Environmental Protection Agency (EPA). [cited 2011 Feb 28];Available from: http://climatechange.supportportal.com/ics/support/default.asp?deptID=23006

17. EPA Victoria. Ecological Footprint - Measuring Our Impact on the Environment [Internet]. State Government of Victoria. [cited 2011 Mar 4];Available from: http://www.epa.vic.gov.au/ecologicalfootprint/default.asp

18. Bernstein L. Climate Change 2007: Synthesis Report [Internet]. Geneva, Switzerland: Intergovernmental Panel on Climate Change (IPCC); 2007. Available from: http://www.ipcc.ch/publications_and_data/publications_ipcc_fourth_assessment_report_synthesis_report.htm

19. US EPA. Greenhouse Gas Emissions from the U.S. Transportation Sector: 1990–2003 [Internet]. Washington, D.C.: US Environmental Protection Agency (EPA); 2006 [cited 2011 Mar 4]. Available from: http://www.epa.gov/oms/climate/420r06003.pdf

20. US EPA. Carbon Sequestration [Internet]. US Environmental Protection Agency (EPA). 2011 Mar 1 [cited 2011 Mar 4];Available from: http://www.epa.gov/r5water/uic/carbon_sequestration.htm

21. US EPA. What is sustainability? [Internet]. US Environmental Protection Agency (EPA). 2011 Feb 27 [cited 2011 Feb 27];Available from: http://www.epa.gov/sustainability/basicinfo.htm

22. Littman T. Sustainable Transportation and TDM [Internet]. Online TDM Encyclopedia: Victoria Transport Policy Institute (VPTI). 2011 Mar 16 [cited 2011 Mar 26];Available from: http://www.vtpi.org/tdm/tdm67.htm

23. United Nations. Kyoto Protocol to the United Nations Framework Convention on Climate Change [Internet]. United Nations (UN); 1998 [cited 2011 Mar 4]. Available from: http://unfccc.int/resource/docs/convkp/kpeng.pdf

24. NOAA. Greenhouse Gases Continue to Climb Despite Economic Slump: Carbon Dioxide, Methane Increased in 2008 [Internet]. National Oceanic and Atmospheric Administration (NOAA). 2009 Apr 21 [cited 2011 Feb 28];Available from: http://www.noaanews.noaa.gov/stories2009/20090421_carbon.html

25. Winder A, Brackstone M, Morin J-M. Thematic ResearchSummary:Road Transport [Internet]. Brussels, Belgium: European Commission Directorate General for Mobility and Transport: Transport Research Knowledge Centre; 2010 [cited 2011 Mar 4]. Available from: http://www.transport-research.info/Upload/Documents/201006/20100608_100042_88270_TRS_Road%20Transport.pdf

26. O'Flaherty C. Highways: the location, design, construction and maintenance of road pavements.

Fourth Edition. China: Butterworth-Heinemann; 2002. 572 p.

27. US EPA. Evaluation of Ecological Impacts from Highway Development [Internet]. Washington, D.C.: US Environmental Protection Agency (EPA); 1994 [cited 2011 Mar 4]. Available from: http://www.epa.gov/compliance/resources/policies/nepa/ecological-impacts-highway-development-pg.pdf

28. Aziz MA. Chapter 3: Environmental Impact Assessment of Highway Development. In: Fwa TF, editor. The handbook of highway engineering. Boca Raton, Florida: CRC Press; 2006. p. 22.

29. Forkenbrock DJ, Weisbrod GE. Guidebook for Assessing the Social and EconomicEffects of Transportation Projects [Internet]. Washington, D.C.: Transportaion Research Board (TRB); 2001 [cited 2011 Mar 25]. Available from: http://onlinepubs.trb.org/onlinepubs/nchrp/nchrp_rpt_456-a.pdf

30. Center for Transportation and the Environment (CTE). Improved Methods for Assessing Social, Cultural, and Economic Effects of Transportation Projects [Internet]. Washington, D.C.: Transportaion Research Board (TRB); 2008 [cited 2011 Mar 26]. Available from: http://www.statewideplanning.org/_resources/234_NCHRP-8-36-66.pdf

31. FHWA. NEPA and Project Development [Internet]. Federal Highway Administration (FHWA). [cited 2011 Mar 26];Available from: http://environment.fhwa.dot.gov/projdev/index.asp

32. UCPRC P-LCA Team. Pavement Life Cycle Assessment Workshop [Internet]. 2010 Sep 8 [cited 2011 Mar 26];Available from: http://www.ucprc.ucdavis.edu/p-lca/resources.html

Chapter 2:
Defining Sustainable Highways and Pavements

Defining Sustainability and Sustainable Development

Over the years, many have attempted to define sustainability and sustainable development, the most popular one being "development that meets the needs of the present without compromising the ability of future generations to meet their own needs" (1). These definitions underline the need for living within the limits and a better understanding of the interconnections among economic growth, environmental quality, and social equity for long-term welfare. It has also been acknowledged that sustainability is not a goal, but a process of continuous improvement according to the needs and the context which can vary in time and space.

In this chapter, we will look at the some of the more well-known definitions related to sustainability, sustainable development, sustainable transportation, and sustainable highway/road pavements.

"Sustainability is a dynamic condition which requires a basic understanding of the interconnections and interdependency among ecological, economic and social systems. Sustainability means providing a rich quality of life for all, and accomplishing this within the means of nature" (2).

"The guiding rules are that people must share with each other and care for the Earth. Humanity must take no more from nature than nature can replenish. This in turn means adopting lifestyles and development paths that respect and work within nature's limits. It can be done without rejecting the many benefits that modern technology has brought, provided that technology also works within those limits" (3).

"A product, process, or service contributes to sustainability if it constrains environmental resource consumption and waste generation to an acceptable level, supports the satisfaction of important human needs, and provides enduring economic value to the business enterprise" (4).

"Sustainable design is the set of perceptual and analytic abilities, ecological wisdom, and practical wherewithal essential to making things that fit in a world of microbes, plants, animals, and entropy. In other words, (sustainable design) is the careful meshing of human purposes with the larger patterns and flows of the natural world, and careful study of those patterns and flows to inform human purposes" (5).

"Sustainable development involves devising a social and economic system, which ensures that these goals are sustained, i.e. that real incomes rise, that educational standards increase, that the health of the nation improves, that the general quality of life is advanced" (6).

"Sustainable development is based on a long-term approach which takes into account the inextricable nature of the environmental, social and economic dimensions of development activities" (7).

"Sustainable development meets the holistic (spiritual and material) needs of the present without comprising the ability of future generations to meet, correspondingly, their own needs" (8).

Defining Sustainable Transportation

There is not a universally accepted definition of sustainable transportation (9). The following are some of the proposed definitions related to sustainable transportation (10,11):

European Conference of Ministers of Transport (ECMT) (12): "A sustainable transport system is one that is accessible, safe, environmentally-friendly, and affordable".

European Commission's Energy, Environment and Sustainable Development Programme (13): "A sustainable urban transport and land use system (a) provides access to goods and services in an efficient way for all inhabitants of the urban area; (b) protects the environment, cultural heritage and ecosystems for the present generation; and (c) does not endanger the opportunities of future generations to reach at least the same welfare level as those living now, including the welfare they derive from their natural environment and cultural heritage."

Transport Canada (TC) (14): "The goal of sustainable transportation is to ensure that environment, social and economic considerations are factored into decisions affecting transportation activity." Some sample goals that may help achieve sustainable transportation depending on a specific context are depicted in Fig. 2.1 (15).

Transportation Research Board (TRB) (16): "...sustainability is not about threat analysis; sustainability is about systems analysis. Specifically, it is about how environmental, economic, and social systems interact to their mutual advantage or disadvantage at various space-based scales of operation."

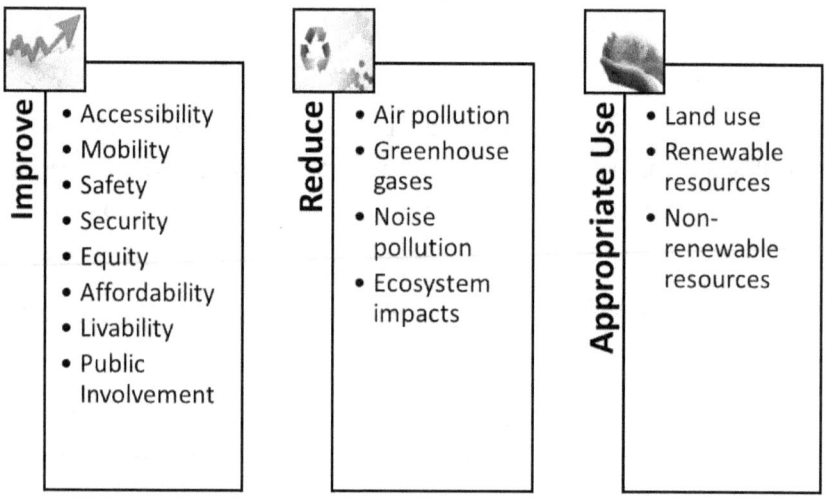

Fig. 2.1. Sample goals for achieving sustainable transportation [*Source:* AASHTO (15)]

Organization for Economic Co-operation and Development (OECD) (17): The OECD defines environmentally sustainable transportation as, "transportation that does not endanger public health or ecosystems and that meets needs for access consistent with (a) use of renewable resources that are below their rates of regeneration, and (b) use of non-renewable resources below the rates of development of renewable substitutes."

Canadian Center for Sustainable Transportation (CST) (18): A sustainable transportation system is one that "(1) allows the basic access and development needs of individuals, companies and society to be met safely and in a manner consistent with human and ecosystem health, and promotes equity within and between successive generations, (2) is affordable, operates fairly and efficiently, offers a choice of transport mode, and supports a competitive economy, as well as balanced regional development, and (3) limits emissions and waste within the planet's ability to absorb them, uses renewable resources at or below their rates of generation, and uses non-renewable resources at or below the rates of development of renewable substitutes, while minimizing the impact on the use of land and the generation of noise."

Transportation Association of Canada (TAC) (19): A sustainable transportation system has the following characteristics:

"(a) In the natural environment: limit emissions and waste (that pollute air, soil and water) within the urban area's ability to absorb/recycle/cleanse, provide power to vehicles from renewable or inexhaustible energy sources. This implies solar power over the long run, and recycle natural resources used in vehicles and infrastructure (such as steel, plastic, etc.);

(b) In society: provide equity of access for people and their goods, in this generation and in all future generations, enhance human health, help support the highest quality of life compatible with available wealth, facilitate urban development at the human scale, limit noise intrusion below levels accepted by communities, and be safe for people and their property, and

(c) In the economy: be financially affordable in each generation, be designed and operated to maximize economic efficiency and minimize economic costs, and help support a strong, vibrant and diverse economy."

New Zealand Ministry for the Environment (20): "Sustainable transport is about finding ways to move people, goods and information in ways that reduce its impact on the environment, the economy, and society."

A sustainable transportation system is "one in which fuel consumption, vehicle emissions, safety, congestion, and social and economic access are of such levels that they can be sustained into the indefinite future without causing great or irreparable harm to future generations of people throughout the world" (21).

Defining Sustainable Highways and Pavements

Green Highways Partnership (GHP) (22): Green highways are those that integrate transportation functionality and ecological sustainability to serve as environmentally responsible and sustainable highways in all aspects, including design, construction, and maintenance. Some physical characteristics of such green highways include biofiltration, recycled pavement materials, wildlife crossing, porous pavement shoulder, etc. (see Fig. 2.2).

Federal Highway Administration (FHWA) (24): "A sustainable highway should satisfy life cycle functional requirements of societal development and economic growth while reducing negative impacts to the environment and consumption of natural resources. The sustainability characteristics of a highway should be assessed and considered from conception through construction and in maintenance and operations throughout its lifecycle".

Another Working Definition by FHWA (25): "Sustainable highways are ones that allow for basic access and development needs in a manner that is consistent with human and ecosystem goals, allow for transport mode and choice regardless of age and ability, are affordable, and operate efficiency – efficiently, limits emissions of – and new and non-renewable resource use. And these principles are equally applicable in urban, suburban or rural areas and can be applied to all functional classifications".

A proposed definition for sustainable highways states, "highway improvements that, from conception to completion, through maintenance & operation satisfy life-cycle functional requirements, while improving the natural, built, social & economic environments *better than before* (added emphasis)" (26).

Sustainability is a characteristic of a roadway project (system) that reflects its capacity to support natural laws (physical constraints or laws of nature – study of ecosystems) and human values (satisfaction of basic human needs and desires – includes the idea of both equity and economy) (27).

USDA Forest Service (28): "An environmentally friendly road should minimize ground disturbance; be well-drained and appropriately surfaced to control erosion and loss of material; employ effective erosion control measures;

and be regularly maintained while continuing to meet user needs."

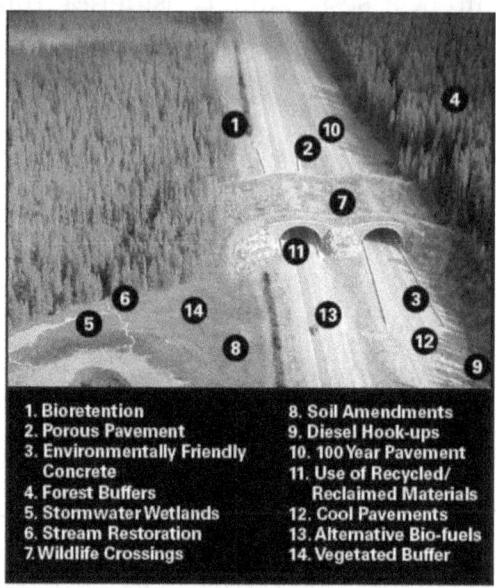

1. Bioretention
2. Porous Pavement
3. Environmentally Friendly Concrete
4. Forest Buffers
5. Stormwater Wetlands
6. Stream Restoration
7. Wildlife Crossings
8. Soil Amendments
9. Diesel Hook-ups
10. 100 Year Pavement
11. Use of Recycled/ Reclaimed Materials
12. Cool Pavements
13. Alternative Bio-fuels
14. Vegetated Buffer

Fig. 2.2. Physical characteristics of green highways with environmental technologies as conceived by Green Highways Partnership [*Source:* GHP (23)]

UK Roads Board (29): "Sustainability in highway maintenance and construction means living within our environmental limits while achieving a sustainable economy".

A sustainable pavement may be defined as "a pavement that minimizes environmental impacts through the reduction of energy consumption, natural resources and associated emissions while meeting all performance conditions and standards" (30).

A sustainable pavement is "is a pavement that is safe and durable, has less maintenance demands, longer time between major rehabilitation interventions, and where life cycle costs are minimized" (31).

A sustainable pavement is a durable pavement that "minimizes the use of energy and renewable resources, while generating a minimum of pollutants, in the most cost-effective manner, while maximizing the benefits to the society" (32).

Sustainable pavement design "must consider life-cycle economic, environmental and societal factors. It is simply good engineering and working with limited resources. It is not about perfection, but instead about balancing competing, and often contradictory, interests" (33).

Useful Online Resources

Sustainable Transportation

Victoria Transport Policy Institute (VTPI): Sustainable Transportation and Transportation Demand Management (TDM) – A chapter that discusses the concepts of sustainability, sustainable development and sustainable transportation, and how TDM can help achieve sustainability goals

http://www.vtpi.org/tdm/tdm67.htm

The Transportation Research Board (TRB) Committee on Transportation and Sustainability website contains lots of information about the committee, including membership, events and activities, and related publications and documents

http://www.trbsustainability.org/

The Center for Environmental Excellence by the American Association of State Highway and Transportation Officials (AASHTO) has been developed in cooperation with the FHWA to promote environmental stewardship and to encourage innovative ways to streamline the transportation delivery process

http://environment.transportation.org/

AASHTO Transportation and Climate Change Resource Center

http://climatechange.transportation.org/

The Canadian Centre for Sustainable Transportation (CST) is a membership-based, non-profit organization that facilitates best practices for the movement of people and goods over the long-term

http://www.centreforsustainabletransportation.org/

The Green Highways Partnership (GHP) serves as a voluntary public-private collaborative that advances environmental stewardship in transportation planning, design, construction, operations and maintenance while balancing economic and social objectives

http://www.greenhighwayspartnership.org/index.php

A list of internet sites related to sustainable transportation

http://www.gdrc.org/uem/sustran/sustran.html

Sustainability and Asphalt Pavements

Asphalt Pavement Alliance (APA)

http://asphaltroads.org/Environment

http://www.quietpavement.com/home.html

National Asphalt Pavement Association (NAPA)

http://www.hotmix.org/index.php?option=com_content&task=view&id=493&Itemid=1058

Asphalt Institute (AI)

http://www.asphaltinstitute.org/

National Center for Asphalt Technology (NCAT)

http://www.ncat.us/

Flexible Pavements of Ohio

http://www.flexiblepavements.org/sustainable_pav.cfm

Asphalt Research Consortium

http://www.arc.unr.edu/

Modified Asphalt Research Center (MARC) at UW-Madison

http://uwmarc.wisc.edu/

Sustainability and Concrete Pavements

American Concrete Pavement Association (ACPA)

http://www.acpa.org/

Portland Cement Association (PCA)

www.cement.org/SD/index.asp

National Concrete Pavement Technology Center (CP Tech Center)

http://www.cptechcenter.org/

National Ready Mixed Concrete Association (NRMCA)

http://www.nrmca.org/sustainability/index.asp

Cement Association of Canada

http://www.cement.ca/en/Concrete-and-the-Environment.html

American Concrete Institute (ACI)

http://www.concrete.org/technical/green-building-resources.htm?HP=b_greenbuilding

Massachusetts Institute of Technology (MIT) Concrete Sustainability Hub

http://web.mit.edu/cshub/

References

1. Brundtland GH. Our Common Future: Report of the World Commission on Enviornment and Development [Internet]. Oslo, Norway: United Nations (UN); 1987 [cited 2011 Feb 27]. Available from: http://www.un-documents.net/wced-ocf.htm

2. Cloud J. Educating for a Sustainable Future [Internet]. In: Jacobs HH, editor. Curriculum 21: essential education for a changing world. ASCD; 2010. Available from: http://www.cloudinstitute.org/brief-history/

3. International Union for Conservation of Nature and Natural Resources (IUCN). Caring for the Earth: A Strategy for Sustainable Living. Earthscan; 2009. 236 p.

4. Fiksel J. Designing resilient, sustainable systems. Environmental science & technology [Internet]. 2003;37(23):5330–5339. Available from:

http://www.resilience.osu.edu/CFR-site/pdf/DesResSusSysFiksel.pdf

5.   Orr DW. Ecological literacy: education and the transition to a postmodern world. SUNY Press; 1992. 240 p.

6.   The University of Reading ECIFM. Definitions of Sustainability [Internet]. [cited 2011 Mar 7];Available from: http://www.ecifm.rdg.ac.uk/definitions.htm

7.   Québec Portal. A definition of sustainable development : Portal of the government of Québec [Internet]. Québec Portal. 2010 Mar 19 [cited 2011 Mar 5];Available from: http://www.gouv.qc.ca/portail/quebec/pgs/commun/go uv/developpementdurable/definitiondevdurable/?lang= en

8.   Cywinski Z. Current Philosophy of Sustainability in Civil Engineering. Journal of Professional Issues in Engineering Education and Practice [Internet]. 2001;127(1):12-16. Available from: http://link.aip.org/link/?QPI/127/12/1

9.   Beatley T. The many meanings of sustainability: Introduction to a special issue of JPL. Journal of Planning Literature. 1995;9(4):339-342.

10. Litman T. Well MeasuredDeveloping Indicators for Sustainable and Livable Transport Planning [Internet]. Canada: Victoria Transport Policy Institute (VTPI); 2011 [cited 2011 Mar 7]. Available from: http://www.vtpi.org/wellmeas.pdf

11. Amekudzi A, Jeon CM. Addressing Sustainability in Transportation Systems: Definitions, Indicators, and Metrics. Journal of Infrastructure Systems [Internet].

2005 [cited 2011 Mar 7];11(31). Available from:
http://ascelibrary.org/iso/resource/1/jitse4/v11/i1/p31_s1

12. ECMT. Assessment & Decision Making for Sustainable
    Transport [Internet]. Paris, France: EUROPEAN
    CONFERENCE OF MINISTERS OF TRANSPORT
    (ECMT); 2004 [cited 2011 Feb 5]. Available from:
    http://www.internationaltransportforum.org/IntOrg/ec
    mt/pubpdf/04Assessment.pdf

13. PROSPECTS. Developing Sustainable Urban Land Use
    and Transport Strategies: A Methodological Guidebook
    [Internet]. Oslo, Norway: Institute of Transport
    Economics; 2003 [cited 2011 Mar 12]. Available from:
    http://www.infra.kth.se/TLA/mtsa/tpe/MethodGuide.p
    df

14. Transport Canada (TC). Sustainable Development -
    Transport Canada [Internet]. 2010 Mar 15 [cited 2011
    Mar 5];Available from:
    http://www.tc.gc.ca/eng/policy/acs-sd-menu.htm

15. AASHTO. Sustainability [Internet]. Center for
    Environmental Excellence by AASHTO. 2011 [cited
    2011 Mar 16];Available from:
    http://environment.transportation.org/environmental_is
    sues/sustainability/

16. TRB. Toward A Sustainable Future: Addressing the
    Long-Term Effects of Motor Vehicle Transportation on
    Climate and Ecology [Internet]. Washington, D.C.:
    Transportaion Research Board (TRB); 1997 [cited 2011
    Mar 5]. Available from:
    http://www.trb.org/Main/Blurbs/Toward_A_Sustainabl
    e_Future_Addressing_the_LongTer_153301.aspx

17. OECD. OECD EnvironmentalIndicators: Towards Sustainable Development [Internet]. Paris, France: Organisation for Economic Co-operation and Development (OECD); 2001. Available from: http://www.oecd.org/dataoecd/37/1/33703867.pdf

18. CST. Definition and Vision of Sustainable Transportation. 2002 Oct;

19. TAC. Acheiving Livable Cities [Internet]. Canada: Transportation Association of Canada (TAC); 1998 [cited 2011 Mar 5]. Available from: http://www.tac-atc.ca/english/resourcecentre/readingroom/pdf/livable.pdf

20. Sustainable Transportation Task Force (STTF). Cross-Canada Survey on Jurisdictions' Approaches and Activities in Sustainable Transportation [Internet]. Ontario, Canada: Council of Ministers Responsible for Transportation and Highway Safety; 2009 [cited 2011 Mar 5]. Available from: http://www.comt.ca/english/crosscanada.pdf

21. Richardson B. Toward a Policy on a Sustainable Transportation System. Transportation Research Record [Internet]. 1999 [cited 2011 Mar 5];1670:27-34. Available from: http://pubsindex.trb.org/view.aspx?id=511933

22. GHP. Green Highways Partnership [Internet]. 2010 [cited 2011 Mar 5];Available from: http://www.greenhighwayspartnership.org/index.php

23. GHP. Fact Sheet FS.01: Collaborating for Progress [Internet]. 2008 [cited 2011 Mar 5];Available from: http://www.greenhighwayspartnership.org/media/GHP FS01COLLABORATINGFORSUCCESS.pdf

24. FHWA. What is a Sustainable Highway? [Internet]. Sustainable Highways Self-Evaluation Tool: Federal Highway Administration. [cited 2011 Mar 6];Available from: https://www.sustainablehighways.org/203/what-is-a-sustainable-highway.html

25. Carlson D. U.S. Department of Transportation (DOT), Federal Highway Administration (FHWA) Sustainability Initiatives [Internet]. 2010 Jan [cited 2011 Mar 8];Available from: http://www.epa.gov/heatisld/resources/pdf/3-Intro-USDOT-FHWA-SustainabilityInitiatives-Carlson-FHWA.pdf

26. Kassoff H. Sustainable Highways: Oxymoron or Opportunity? [Internet]. 2006 Nov 29;Available from: http://trid.trb.org/view.aspx?id=626073

27. Anderson JL. Sustainability in Civil Engineering [M.S. Thesis]. [Seattle, WA]: University of Washington; 2008. 336 p.

28. Keller G, Sherar J. Low-Volume Roads Engineering: Best Management Practices Field Guide [Internet]. Washington, D.C.: USDA Forest Service; 2003. Available from: http://ntl.bts.gov/lib/24000/24600/24650/Index_BMP_Field_Guide.htm

29. Reid JM, Chandler JWE, Schiavi I, Hewitt AP. Sustainable Highways: A Short Guide [Internet]. London, UK: Department for Transport (UK Roads Board); 2008 [cited 2011 Mar 14]. Available from: http://www.handapp.co.uk/web/Sustainable%20highways%20A%20short%20guide.pdf

30. Miller TD, Bahia HU. Sustainable Asphalt Pavements: Technologies, Knowledge Gaps and Opportunities - A White Paper [Internet]. 2009 Feb [cited 2011 Feb 2];Available from: http://uwmarc.wisc.edu/files/MARC-Sustainable-Asphalt-Pavements-white-paper.pdf

31. Maher MLJ, Uzarowski L, Moore G. Designing High Traffic Volume Urban Motorway Pavements to Maximize Sustainability [Internet]. In: Rauch S, Morrison GM, Monzón A, editors. Highway and Urban Environment. Netherlands: Springer; 2010. p. 367-376.Available from: http://dx.doi.org/10.1007/978-90-481-3043-6_39

32. Taylor P. Two-Lift Construction: KS Demonstration Project [Internet]. 2008 Oct 15 [cited 2011 Mar 5];Available from: http://www.cptechcenter.org/projects/two-lift-paving/documents/IntrotoTwo-Lift-Taylor.pdf

33. Van Dam T. Quantifying Environmental Impact of Concrete Pavements: What Should We Measure and How Is It Done? [Internet]. 2009 Aug [cited 2011 Mar 16];Available from: http://www.appliedpavement.com/presentations/tvd/Measurement_tvd_aug09x.pdf

Chapter 3:
Sustainable Transportation Initiatives

OECD's Environmentally Sustainable Transport (EST)

The topic of sustainable transportation has been gaining increasing attention over the past decade with continued growth in the number of motorized vehicles, their use, and their negative impact on the environment and ecosystem. Several initiatives have been initiated across the globe to make transportation more sustainable (1).

Recognizing the need for a new policy approach which places environmental criteria in the front along with other policy goals, the Organisation for Economic Co-operation and Development (OECD) Environmental Policy Committee's Task Force on Transport initiated the Environmentally Sustainable Transport (EST) project in 1994 with the objectives of developing methods and policy guidelines for sustainable transportation (1).

The EST approach uses a backcasting exercise. In the first step, different "what-if" EST scenarios are set out. In the second step, EST is characterized in terms of quantifiable transportation activity targets and the environmental impacts of each activity unit. The third step involves determining the actions required for meeting the targets by working back from these targets to present conditions (3).

According to OECD EST Guidelines endorsed at an international conference held in 2000, the concept of EST is defined as one where "transportation does not endanger public health or ecosystems and meets needs for access consistent with (a) use of renewable resources below their rates of regeneration, and (b) use of nonrenewable resources below the rates of development of renewable substitutes"(3).

The UNCRD Asian EST Initiative

The United Nations Centre for Regional Development's (UNCRD's) Asian EST Initiative was inspired by the OECD's EST Initiative, but with considerations also given to the different levels of economic development between OECD member countries and developing countries in Asia. In the August of 2005, the *Aichi Statement* was adopted by the participants of the International Conference on Environment and Transport, Nagoya City, Aichi Prefecture, Japan towards establishment of the a Regional EST Forum for the promotion of environmentally sustainable transport in Asia.

The *Aichi Statement* proposed an integrated approach to EST which included public health, environment and people friendly urban transport infrastructures, public transport planning and transport demand management (TDM), social equity and gender perspectives, etc (4).

Complete Streets

A broad coalition of advocates and transportation professionals, under the banner of The National Complete Streets Coalition, have been working together to promote policy and procedural changes (referred to as Complete Streets) at the Federal, State and local levels across the United States to build road networks that are safer, more livable, and welcoming to everyone, and not just the automobile traffic. The coalition includes the American Public Transportation Association, the American Planning Association, AARP, Institution of Transportation Engineers (ITE) and many others (5).

A Complete Streets policy will ensure that an entire roadway will be designed and operated with all users in mind, including bicyclists, public transportation vehicles and riders, and pedestrians of all ages and abilities. Although the FHWA does not have an official Complete Streets policy, The Complete Streets policy is consistent with the FHWA's design guidance on Accommodating Bicycle and Pedestrian Travel: A Recommended Approach that "bicycling and walking facilities will be incorporated into all transportation projects unless exceptional circumstances exist" (6).

Previously, street design in the U.S. was treated as a relatively straightforward task based on roadway functional classification schemes oriented toward automobile mobility. This often resulted in sterile, inhospitable streetscapes in developing suburbs (7). In recent years, with the development of CSD and CSS concepts and the adoption of Complete Streets policy by transportation agencies, the focus has been shifting towards consideration of various user perspectives (bicyclists, pedestrians, etc.) and the surrounding land use context, in addition to the street function.

Although there is no universal Complete Street design and each one will need to respond to the local community's context, the following are some of the elements a Complete Street may include according to the National Complete Streets Coalition (see Fig. 3.1): sidewalks, bike lanes (or wide paved shoulders), special bus lanes, comfortable and accessible public transportation stops, frequent and safe crossing opportunities, median islands, accessible pedestrian signals, curb extensions, narrower travel lanes, roundabouts, and more (8).

Fig. 3.1. Some elements of a "Complete Street" [*Source: National Complete Streets Coalition* (5)]

The Complete Streets movement has been growing in popularity over the last few years. As of January 2011, over 200 jurisdictions across the U.S. have formally committed to adopt Complete Streets policies or some form of it. The

American Academy of Pediatrics and the Centers for Disease Control (CDC) and Prevention named adoption of Complete Streets policies as a recommended strategy for preventing obesity, and improving safety and increasing physical activity among children and adults (5).

Similar to the Complete Streets policy, the US DOT provided the following policy statement (dated March 15, 2010) to reflect the DOT's support for the development of fully integrated transportation networks: "The DOT policy is to incorporate safe and convenient walking and bicycling facilities into transportation projects. Every transportation agency, including DOT, has the responsibility to improve conditions and opportunities for walking and bicycling and to integrate walking and bicycling into their transportation systems. Because of the numerous individual and community benefits that walking and bicycling provide — including health, safety, environmental, transportation, and quality of life — transportation agencies are encouraged to go beyond minimum standards to provide safe and convenient facilities for these modes" (9). An example of a Complete Streets planning process included in the Charlotte's Urban Streets Design Guidelines is shown in Fig. 3.2 (10).

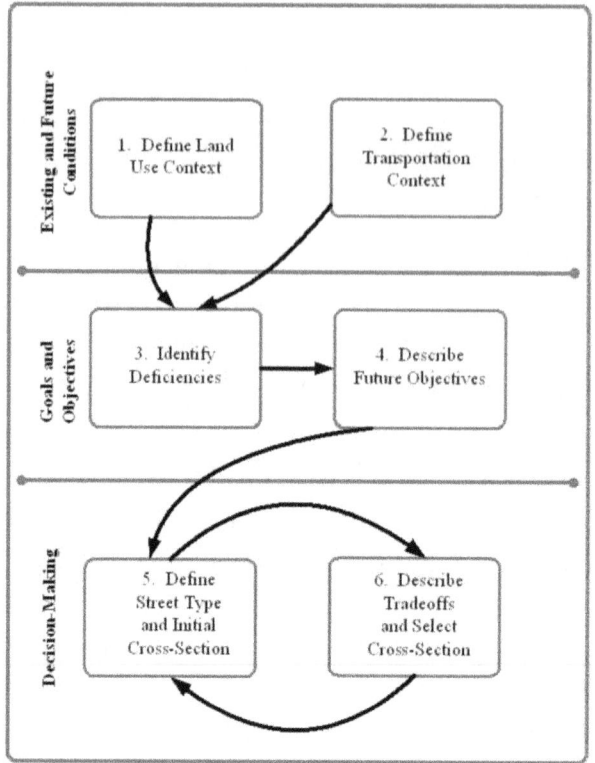

Fig. 3.2. An example of a Complete Streets planning process in applying Charlotte's urban street design guidelines [*Source:* City of Charlotte (10)]

Context Sensitive Solutions (CSS)

Similar to the Complete Streets initiative, CSS is "a collaborative, interdisciplinary approach that involves all stakeholders to develop a transportation facility that fits its physical setting and preserves scenic, aesthetic, historic and environmental resources, while maintaining safety and mobility"(11). Unlike the traditional approach where technical experts typically developed transportation projects and presented to the public after the majority of decisions were already made, the CSS approach simultaneously engages community stakeholders and interdisciplinary teams to resolve transportation problems together (see Fig. 3.3) (12).

The CSS approach to transportation decision-making and design takes into account "the context" (i.e. communities and lands) through which streets, roads, and highways pass and involves the public and stakeholders throughout the project development process right from the beginning. An example of urban context is shown in Fig. 3.4 which includes built and natural environmental as well as social, cultural, and economic aspects beyond the function and design of the transportation facility. According to the FHWA, "CSS is an approach that considers the total context within which a transportation improvement project will exist"(13).

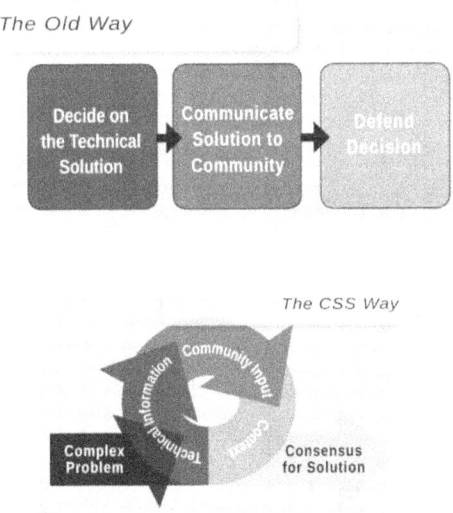

Fig. 3.3. The Context Sensitive Solutions (CSS) approach to project development contrasted with the traditional approach [Source: FHWA (12)]

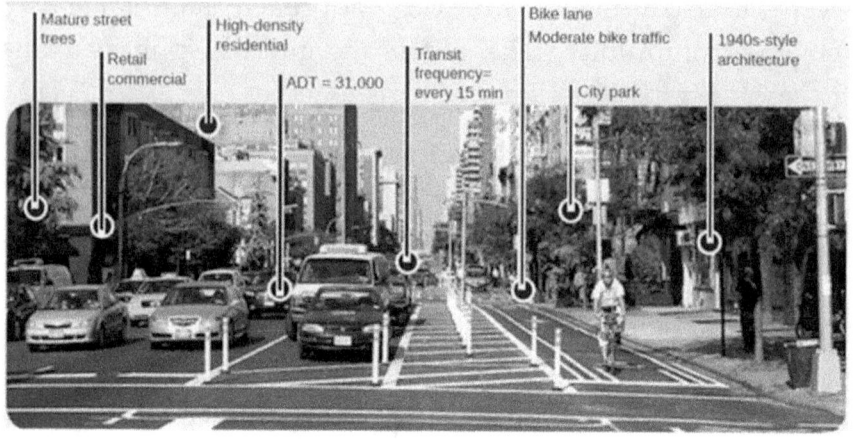

Fig. 3.4. Example of urban context in Context Sensitive Solutions (CSS) [*Source:* FHWA (12)]

CSS seeks to balance the need for safe and efficient movement of vehicles with other desirable outcomes, including environmental sustainability and historic preservation. Context Sensitive Design (CSD) is a concept that is similar to CSS, with a slightly different emphasis. CSD is restricted to the design phase whereas CSS extends to all decisions in transportation planning, development, operations and maintenance (14).

CSS draws its roots from the US National Environmental Policy Act (NEPA) of 1969 which set forth a framework of objectives to facilitate environmental planning and decision-making by Federal agencies. These objectives include environmental protection, interagency coordination and cooperation, and public participation in planning and project development, which later on became the key elements of CSS (15).

The FHWA has been promoting the CSD/CSS philosophy with a commitment to advance CSS nationwide as one of the objectives of its Vital Few Goal on Environmental Stewardship and Streamlining: "to improve the environmental quality of transportation decision making by incorporating context sensitive solutions principles in all aspects of planning and the project development process"(13).

In collaboration with Project for Public Spaces, Scenic America, AASHTO, Federal Transit Administration, Institute for Transportation Engineers (ITE), National Association of City Transportation Officials (NACTO), and National Park Service (NPS), the FHWA also sponsors a Clearinghouse (website) to serve as a resource for state and local transportation agencies and practitioners, and community stakeholders (13).

Green Highways Partnership (GHP)

The GHP serves as a voluntary public-privative initiative that advances environmental stewardship in transportation planning, design, construction, operations, and maintenance (16). GHP was launched in 2005 through a coordinated effort by the US EPA, FHWA, TRB, AASHTO and private transportation consultants to leverage partnerships, recognition, and opportunities to promote, develop, and demonstrate sustainable transportation solutions (16).

According to GHP, green highways integrate transportation functionality and ecological sustainability and can include any combination of the characteristics shown in Fig. 3.5 depending on the project and location.

The GHP partnership action themes primarily focuses on watershed-driven stormwater management; recycling and reuse (industrial by-products uses/implementation); and conservation and ecosystem management and also integrates the goals of existing initiatives such as CSS, environmental stewardship, and sustainability (16).

FHWA Eco-Logical

A steering team comprised of Federal agencies and State DOTs developed Eco-Logical, a framework enabling partners involved in infrastructure development to work together and make it more sensitive to wildlife and their ecosystems. Integrated planning, the exploration of a variety of mitigation options, and performance measurement are some key components of the Eco-Logical guide which "describes ways for ways for streamlining the processes that advance approvals for infrastructure projects – in compliance with applicable laws – while maintaining safety, environmental health, and effective public involvement" .

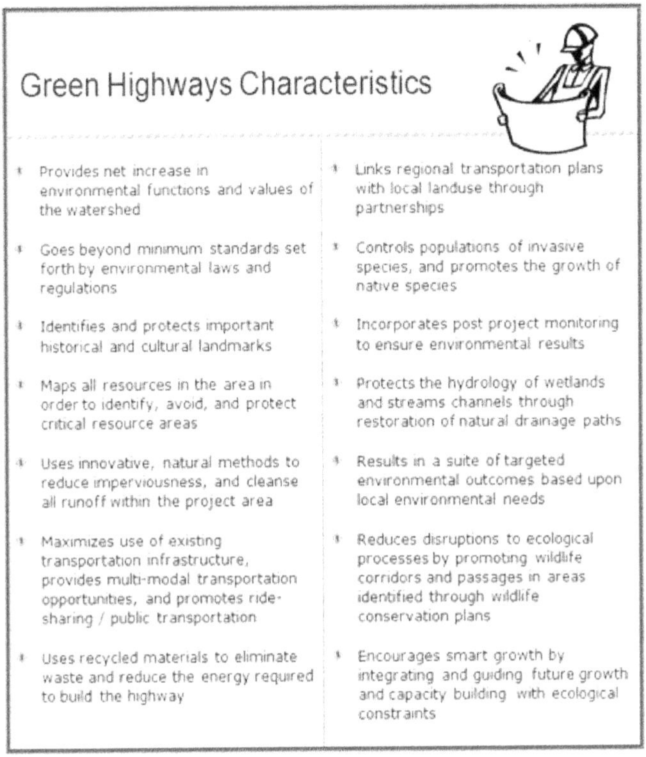

Green Highways Characteristics

- Provides net increase in environmental functions and values of the watershed

- Goes beyond minimum standards set forth by environmental laws and regulations

- Identifies and protects important historical and cultural landmarks

- Maps all resources in the area in order to identify, avoid, and protect critical resource areas

- Uses innovative, natural methods to reduce imperviousness, and cleanse all runoff within the project area

- Maximizes use of existing transportation infrastructure, provides multi-modal transportation opportunities, and promotes ride-sharing / public transportation

- Uses recycled materials to eliminate waste and reduce the energy required to build the highway

- Links regional transportation plans with local landuse through partnerships

- Controls populations of invasive species, and promotes the growth of native species

- Incorporates post project monitoring to ensure environmental results

- Protects the hydrology of wetlands and streams channels through restoration of natural drainage paths

- Results in a suite of targeted environmental outcomes based upon local environmental needs

- Reduces disruptions to ecological processes by promoting wildlife corridors and passages in areas identified through wildlife conservation plans

- Encourages smart growth by integrating and guiding future growth and capacity building with ecological constraints

Fig. 3.5. Characteristics of green highways according to Green Highways Partnership [*Source:* GHP (16)]

The expected benefits of an ecosystem approach to infrastructure projects include (17,18):

✓ Safer, improved infrastructure

✓ Improved watershed and ecosystem health

✓ Increased connectivity and conservation

✓ Efficient project development

✓ Increased transparency

FHWA Exemplary Ecosystem Initiatives (EEIs)

According to FHWA, "an EEI is an action or measure that will help sustain or restore natural ecosystems and their functions and values, using an ecosystem or landscape approach"(19). To demonstrate its commitment to ecosystem and habitat conservation, the FHWA originally agreed to identify a minimum of 30 EEIs in at least 20 States or Federal Lands Highway divisions by September 2007 (20). Examples of such initiatives include "mitigation projects that support wildlife movement and habitat connectivity, maintain ecosystem integrity, implement watershed-based environmental assessment and mitigation approaches, and encourage the use of wetland and habitat banking and the use of special measures to prevent invasive species along highway rights-of-way"(19). Recent EEIs trends and targets are shown in Fig. 3.6 (19).

The following criteria were formulated by the FHWA to aid in identifying and selecting the initiatives (21):

✓ "An EEI helps sustain or restore natural systems and their functions and values.

✓ An EEI is developed within a landscape context.

✓ An EEI uses partnering and collaborative approaches to advance common goals.

✓ An EEI uses the best available science in ecosystem and habitat conservation.

✓ An EEI provides clear examples of innovative environmental solutions by transportation agencies and achieves high standards in the environmental process.

✓ An EEI achieves high quality results.

✓ An EEI is recognized by environmental interests as being particularly valuable or noteworthy".

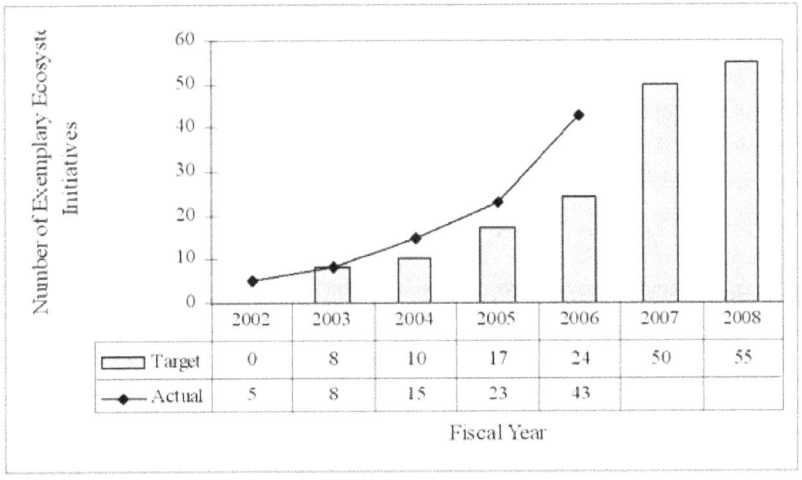

	2002	2003	2004	2005	2006	2007	2008
Target	0	8	10	17	24	50	55
Actual	5	8	15	23	43		

Fiscal Year

Fig. 3.6. FHWA Exemplary Ecosystem Initiatives, Fiscal Year 2002 to 2008 [Source: FHWA (19)]

FHWA Exemplary Human Environment Initiatives (EHEIs)

The purpose of FHWA's EHEIs (which began in 2007) is to recognize "outstanding examples of transportation projects that either create or improve conditions for human activities while protecting the natural environment" (22). The goal of EHEIs is "to make things better for people when they use our Nation's transportation system while remaining conscious of any natural environmental consequences" (22).

Since 2008, the FHWA has been allowing projects to be submitted for joint consideration with the EEIs recognizing that environmental achievements do not have to be divided between natural and human initiatives. The EHEIs award categories include (22):

✓ Encouraging Nonmotorized Transportation

✓ Enhancing the Environment for Human Activities

✓ Process Improvements

✓ Educational and Training Programs

✓ Product Development

✓ Others

FHWA Livability Initiative

Recognizing the multiple co-benefits associated with transportation investments that support community livability, the FHWA began the Livability initiative to "enhance the environmental sensitivity of roads and bridges and to help states explore multi-modal transportation options" (23). Livability in transportation is not a new concept. It has been advocated by communities, developers, businesses, and neighborhood residents for many generations using a range of names with overlapping set of objectives and strategies: livability, sustainability, smart growth, walkable communities, new urbanism, healthy neighborhoods, active living, transit-oriented development (TOD), Complete Streets, etc (24).

According to FHWA, "Livability is about tying the quality and location of transportation facilities to broader opportunities such as access to good jobs, affordable housing, quality schools, and safe streets. This includes addressing safety and capacity issues on all roads through better planning and design, maximizing and expanding new technologies such as ITS and the use of quiet pavements, using Travel Demand Management approaches to system planning and operations, etc." (23). The Livability principles include:

✓ Provide more transportation choices

✓ Promote equitable, affordable housing

✓ Enhance economic competitiveness

✓ Support existing communities

✓ Coordinate and leverage Federal policies and investment

✓ Value communities and neighborhoods

Under the FHWA's Livability Initiative, the Partnership for Sustainable Communities was formed by the U.S. Department of Housing and Urban Development (HUD), the U.S. Department of Transportation (DOT), and the U.S. Environmental Protection Agency (EPA) in June 2009 "to ensure that housing and transportation goals are met while simultaneously protecting the environment, promoting equitable development, and helping to address the challenges of climate change" (25).

The principles behind HUD-DOT-EPA Partnership for Sustainable Communities are shown in Fig. 3.7 (26). The HUD/DOT/EPA partnership is expected to bestow the following benefits (27):

✓ Enhance integrated planning and investment

✓ Provide a vision for sustainable growth

✓ Redefine housing affordability and make it transparent

✓ Redevelop underutilized sites

✓ Develop livability measures and tools

✓ Align HUD, DOT, and EPA programs

✓ Undertake joint research, data collection, and outreach

Provide More Transportation Choices   **Coordinate Policies and Leverage Investment**   **Promote Equitable, Affordable Housing**

Enhance Economic Competitiveness   **Support Existing Communities**   **Value Communities and Neighborhoods**

Fig. 3.7. Principles of HUD-DOT-EPA Sustainable Communities Partnership [*Source:* FHWA (26)]

French Ministry of Ecology, Sustainable Development, Transport and Housing

An Environment Round Table, named "Grenelle-Environment", was initiated by the French Government in 2007-2008 to define key points of the government policy on ecological and sustainable development for the upcoming years. The State, unions, employers, NGOs and local authorities were brought together as representatives of civilian and public services representatives at the discussion table (28). The key points include:

✓ A fight against climate change

✓ Preserve biodiversity and natural resources

✓ Prevent environmental risks

✓ Promote environmental development patterns favorable to competitiveness and employment

In the road sector, three national federations of public works and earthworks companies along with The French Ministry of Ecology, Energy, Sustainable Development and the Sea and with the Association of the local Authorities (Assemblée des Départements de France) as well as the federation of engineering companies signed a voluntary agreement on March 25, 2009 towards achieving sustainable construction and maintenance of road infrastructures. Sétra, the Technical Department for Transport, Roads and Bridges Engineering (Service d'études sur les transports, les routes et leurs aménagements) of the French Ministry of Ecology, Sustainable Development, Transport and Housing led the discussions and negotiations with the federations (29). The main points of this agreement are listed below (28):

✓ Preserve non-renewable resources

✓ Preserve biodiversity and natural environment

✓ Reduce GHG emissions and energy consumption

✓ Reduce consumption of water and on earthwork sites

✓ Increase the environmental performance of road companies

✓ Improve road safety, workers, users and local residents safety during roadworks

✓ Participate in the development of research and dissemination of innovation

✓ Create a French Institute of Roads and Transport Infrastructures to create conditions for genuine ecological competitiveness

Useful Online Resources

Organisation for Economic Co-operation and Development (OECD): Transport and Environment

http://www.oecd.org/department/0,3355,en_2649_34363_1_1_1_1_1,00.html

National Complete Streets Coalition

http://www.completestreets.org/

Context Sensitive Solutions (CSS)

http://contextsensitivesolutions.org/

FHWA and Context Sensitive Solutions

http://www.fhwa.dot.gov/context/index.cfm

US HUD-DOT-EPA Partnership for Sustainable Communities

http://www.epa.gov/smartgrowth/partnership/

FHWA Eco-Logical

http://www.environment.fhwa.dot.gov/ecological/eco_index.asp

FHWA Exemplary Human Environment Initiatives (EHEI)

http://www.fhwa.dot.gov/environment/ehei/index.htm

FHWA Livability Initiative

http://www.fhwa.dot.gov/livability/

VTPI TDM Encyclopedia: Community Livability

http://www.vtpi.org/tdm/tdm97.htm

References

1. OECD. Indicators for the Integration of Environmental Concerns into Transport Policies [Internet]. Organisation for Economic Co-operation and Development (OECD); 1999 [cited 2011 Mar 8]. Available from: http://www.oecd.org/officialdocuments/displaydocume ntpdf?cote=ENV/EPOC/SE%2898%291/FINAL&doclang uage=en

2. ECMT. Assessment & Decision Making for Sustainable Transport [Internet]. Paris, France: EUROPEAN CONFERENCE OF MINISTERS OF TRANSPORT (ECMT); 2004 [cited 2011 Feb 5]. Available from: http://www.internationaltransportforum.org/IntOrg/ec mt/pubpdf/04Assessment.pdf

3. OECD. OECD Guidelines towards Environmentally Sustainable Transport [Internet]. Paris, France: Organisation for Economic Co-operation and Development (OECD); 2002 [cited 2011 Mar 8]. Available from: http://esteast.unep.ch/phocadownload/oecd9714.pdf

4. Onogawa K. The UNCRD Asian ESTInitiative [Internet]. Guadalajara, Mexico: United Nations Centre for Regional Development (UNCRD); 2008 [cited 2011 Mar 8]. Available from: http://www.oecd.org/dataoecd/46/28/41504031.pdf

5. National Complete Streets Coalition. Complete Streets [Internet]. 2010 [cited 2011 Mar 11];Available from: http://www.completestreets.org/

6. FHWA. Design Guidance - Accommodating Bicycle and Pedestrian Travel: A Recommended Approach - A US DOT Policy Statement [Internet]. Bicycle and Pedestrian Design Guidance. 2008 Mar 26 [cited 2011 Mar 11];Available from: http://www.fhwa.dot.gov/environment/bikeped/design.htm

7. Laplante J, McCann B. Complete Streets: We Can Get There from Here. ITE JOURNAL [Internet]. 2008 [cited 2011 Mar 11];78(5):24-28. Available from: http://www.urbandesignforum.org.nz/PortalDocs/UDF/pdf/CompleteStreets_ITEMay2008.pdf

8. Smith R, Reed S, Baker S. Street Design: Part 1— Complete Streets. Public Roads [Internet]. 2010 Aug [cited 2011 Mar 11];74(1). Available from: http://www.fhwa.dot.gov/publications/publicroads/10julaug/03.cfm

9. US DOT. United States Department of Transportation Policy Statement on Bicycle and Pedestrian Accommodation Regulations and Recommendations [Internet]. Bicycle and Pedestrian Guidance - FHWA. 2010 Mar 15 [cited 2011 Mar 11];Available from: http://www.fhwa.dot.gov/environment/bikeped/policy_accom.htm

10. City of Charlotte. Urban Street Design Guidelines (USDG) [Internet]. 2010 [cited 2011 Mar 11];Available from: http://charmeck.org/city/charlotte/Transportation/Plans Projects/Pages/Urban%20Street%20Design%20Guidelines.aspx

11. Context Sensitive Solutions. What is CSS? [Internet]. 2005 [cited 2011 Mar 10];Available from: http://contextsensitivesolutions.org/content/topics/what _is_css/

12. FHWA. Context Sensitive Solutions Primer: In Support of Livable Communities and Sustainable Transportation [Internet]. Washington, D.C.: Federal Highway Administration (FHWA); [cited 2011 Mar 11]. Available from: http://www.fhwa.dot.gov/context/css_primer/docs/FH WA_CSS_Primer.pdf

13. FHWA. FHWA and Context Sensitive Solutions (CSS) [Internet]. 2010 Dec [cited 2011 Mar 10];Available from: http://www.fhwa.dot.gov/context/index.cfm

14. Victoria Transport Policy Institute (VTPI). Context Sensitive Design: Roadway Design That Is Responsive To Local Community Values [Internet]. 2010 Nov [cited 2011 Mar 10];Available from: http://www.vtpi.org/tdm/tdm57.htm

15. FHWA Resource Center. CSS Quick Facts – How CSS Developed [Internet]. [cited 2011 Mar 10];Available from: http://contextsensitivesolutions.org/content/reading/ho w_css_developed___css_quick_f_/resources/How_CSS_ Developed_Quick_Facts.pdf/

16. Green Highways Partnership (GHP). About GHP [Internet]. 2010 [cited 2011 Mar 10];Available from: http://www.greenhighwayspartnership.org/index.php? option=com_content&view=article&id=2&Itemid=29#W HATARESTEPS

17. FHWA. FHWA Eco-Logical: An Ecosystem Approach to Developing Infrastructure Projects [Internet]. FHWA, U.S. Department of Transportation. 2006 [cited 2011 Mar 11];Available from: http://environment.fhwa.dot.gov/ecological/eco_index.a sp

18. Brown JW. Eco-logical: An Ecosystem Approach to Developing Infrastructure Projects [Internet]. Washington, D.C.: Federal Highway Administration (FHWA); 2006 [cited 2011 Mar 11]. Available from: http://environment.fhwa.dot.gov/ecological/ecological.p df

19. FHWA. FHWA Budget Estimates Fiscal Year 2008 [Internet]. Section 4: Performance Budget - FHWA Budget Estimates Fiscal Year 2008. 2008 [cited 2011 Mar 12];Available from: http://www.fhwa.dot.gov/budget/fy2008/section4.htm

20. FHWA. Exemplary Ecosystems Initiatives [Internet]. FHWA, U.S. Department of Transportation. 2010 [cited 2011 Mar 11];Available from: http://www.environment.fhwa.dot.gov/ecosystems/eei. asp

21. Shrouds JM. Memorandum: Exemplary Ecosystem Initiatives Criteria [Internet]. 2003 Oct 6 [cited 2011 Mar 11];Available from: http://www.environment.fhwa.dot.gov/ecosystems/eei/ ecoinitm.asp

22. FHWA. Exemplary Human Environment Initiatives (EHEI) [Internet]. FHWA, U.S. Department of Transportation. 2010 Aug 4 [cited 2011 Mar

11];Available from:
http://www.fhwa.dot.gov/environment/ehei/index.htm

23. FHWA. Livability Initiative [Internet]. Livability - FHWA. 2010 [cited 2011 Mar 11];Available from: http://www.fhwa.dot.gov/livability/

24. ICF International and AECOM. Livability in Transportation Guidebook: Planning Approaches that Promote Livability [Internet]. Washington, D.C.: Federal Highway Administration (FHWA); 2010 [cited 2011 Mar 11]. Available from: http://www.fhwa.dot.gov/livability/case_studies/guideb ook/livabilitygb10.pdf

25. FHWA. Leveraging the Partnership: DOT, HUD, and EPA Programs for Sustainable Communities [Internet]. FHWA Livability. 2009 [cited 2011 Mar 11];Available from: http://www.fhwa.dot.gov/livability/scp.cfm

26. FHWA. STEP & Livability - Surface Transportation Environment and Planning Cooperative Research Program (STEP) [Internet]. 2010 May 26 [cited 2011 Mar 12];Available from: http://www.fhwa.dot.gov/hep/step/may_26_2010_webi nar/index.htm

27. US EPA. HUD-DOT-EPA Partnership for Sustainable Communities - Smart Growth [Internet]. US Environmental Protection Agency (EPA). 2011 Feb 3 [cited 2011 Mar 11];Available from: http://www.epa.gov/smartgrowth/partnership/index.ht ml

28. Danzanvilliers P. Building Sustainable Roads: A Voluntary Commitment of Road Construction and

Maintenance Companies [Internet]. In: Proceedings of European Transport Conference (ETC) 2009. Glasgow, Scotland, UK: Association for European Transport (AET); 2009 [cited 2011 Mar 27]. Available from: http://www.etcproceedings.org/paper/download/3908

29. Sétra. About Sétra [Internet]. Sétra: Technical Department for Transport, Roads and Bridges Engineering and Road Safety of the french Ministry of Ecology, Sustainable Development, Transport and Housing. 2006 [cited 2011 Mar 27];Available from: http://www.setra.equipement.gouv.fr/English-presentation.html

Chapter 4:
Environmental, Economic and Social Impacts of Highways and Pavements

In this chapter, we will consider the overall sustainability impacts associated with highways and pavements throughout their life, from "cradle" to "grave". Then, we will discuss the commonly used and existing frameworks for quantifying environmental sustainability of products or processes from a life-cycle perspective, the different environmental impact assessment tools developed for evaluating highway and pavement projects, and the relevant literature.

First, we will consider only the negative social, economic, and environmental impacts resulting from the following life cycle activities of highways and pavements:

✓ Materials Production Phase
- Extraction of raw materials
- Pavement materials production (cement, asphalt, etc.)
- Mixing process of HMA or PCC in plants
- Transport of materials between facilities

✓ **Construction, Maintenance & Rehabilitation Phase**

- Transport of materials and construction/maintenance equipment to site

- Use of onsite construction/maintenance equipment

- Traffic delay caused by construction activities and the associated emission/fuel consumption

- Water transport and use

- Storm water system (drainage)

- Building of roadway lighting system

✓ **Operational or Use Phase**

- Vehicle rolling resistance and fuel economy

- Damage to vehicles and freight from pavement deterioration

- Tire wear

- Roadway lighting

- Heat island effect

- Carbonation of concrete

- Leachate and runoff

✓ **End of Life Phase**

- Demolition

- Disposal in a landfill

- Recycling processes

Materials Production Phase Impacts

Extraction of Materials from Quarries

The production of materials used in roadway construction accounts for 60% to 80% of energy used during construction and produces 60% to 90% of CO_2 emissions associated with construction (1). The extraction of primary materials from quarries and borrow pits for road construction not only depletes non-renewable resources, but has associated environmental impacts such as loss of mature countryside, change in ecosystem services, visual intrusion, noise, dust, and blasting vibration (see Fig. 4.1). Additionally, wastewater containing suspended solids and colloidal fines is a major source of hydrospheric emissions from quarrying operations (2).

About 95% of asphalt pavements and about 80% of concrete pavements are made up of aggregates. The 4 million miles of roadway in the US (as of 2006) is estimated to include 1.5 billion metric tons of natural (virgin) aggregates and 6 million metric tons of steel (3). In 2010, 50% of the aggregates were produced in only 10 states in the US due to resource depletion, zoning regulations, and alternative land uses restricting mining (4). In Denmark, approximately 65% of total volume of natural aggregates (more than 1 billion ft^3) extracted annually is used in the road sector (5).

Fig. 4.1. Suzio York Hill crushed stone quarry located near Meriden, Connecticut [*Source:* USGS (6)]

Asphalt and Cement Production

The US alone consumes 35 million metric tons of asphalt and 48 million metric tons of concrete annually in its transportation infrastructure system, at a cost of nearly $65 billion (1). A significant amount of energy is required for producing asphalt binder from crude petroleum and for drying aggregates. For instance, to dry and heat just one ton of aggregates used in road construction, approximately 1.5 gallons of fossil fuel is used. Asphalt contains 7,354 MJ of feedstock energy per cubic meter (7). Asphalt is also a large source of volatile organic compounds (VOC) accounting for

200,000 metric tons of VOC emissions annually from asphalt pavement construction (8).

Similarly, the worldwide production of cement releases more than 1.6 billion metric tons of CO_2 annually which accounts for over 8% of total CO_2 emissions from all human activities (9). During the production of cement, significant levels of other pollutants, such as particulate matter and sulfur oxides are also released (9).

Construction, Maintenance & Rehabilitation Phase Impacts

The road construction industry is a significant contributor to the socio-economic development of any nation. Unfortunately, its negative impacts on environment and human health is also significant. The construction industry has been held responsible for high-energy consumption, solid waste generation, GHG emissions, pollution, and environmental damage, and resource depletion (10). Because of heavy initial processing of raw materials in pavement construction, the amount of primary energy requirement and the associated CO_2 emissions are relatively higher compared to building construction (11).

Land Use

In terms of land use, roughly, 1.5 to 2.0% of the world's total land surface is devoted mainly for roads and parking lots and about 30 to 60% urban areas are covered by highway infrastructure alone (2). An estimated 34,500 square miles in the US, an area roughly equal to the State of Illinois, are covered by pavement surfaces (12).

In the US, the transportation infrastructure was valued at $2.97 trillion or 32% of the value of all fixed assets in 2008. The Federal, State, and local governments own 92% of the infrastructure and the rest is privately owned. Every year, the US invests $100 billion in roadway construction (see Fig. 4.2). In some countries like India where more than 80% of passenger traffic commute via roads, the Government invests approximately $26 billion in just building the road network across the country (13).

The impacts of road construction work is not always limited to the construction zone vicinity, but may result in roadway users using detours around the construction site which will have an impact on the surrounding land use such as noise and vibrations. In addition, there are potential safety concerns of re-routed traffic resulting in higher traffic volumes through residential areas that are not normally trafficked (14).

The sources of construction noise which cause annoyance to the people living adjacent to the roadway include on-site construction vehicles transporting construction vehicles, use of construction equipment such as excavators, compactors, pile drivers, generators, vibrators, etc (14). In addition, the lack of roadway availability during construction activities can negatively influence the mobility of the traveling public, result in additional cost to the traveling public due to traffic delays, and have economic impact on local businesses (14).

Fig. 4.2. Every year the US invests $100 billion in roadway construction [*Source*: FHWA, NYSDOT (15)]

Habitat Fragmentation

To build transportation corridors, huge tracts of land are cleared removing these acres from constructive uses and resulting in deforestation. One of the major ecological impacts of transportation infrastructure has been habitat fragmentation, i.e., splitting of natural habitats and ecosystems into smaller, more isolated patches (see Fig. 4.3).

In 1996, the length of the Trans-European Transport Network covered 1.2% of the total land area. The length of highways planned for construction across Europe in just 2010 is more than 12,000 miles. Expansion of transportation infrastructure networks reduces the suitability of adjacent areas for wildlife and creating barriers to animal movement

(16). Over the past 15 years, collisions between wildlife and vehicles have increased by 50 percent, costing the US $8 billion annually (17).

Fig. 4.3. One of the major ecological impacts of transportation infrastructure is habitat fragmentation [*Source:* USGS (left); FHWA (right)]

Energy and Water Use

It is estimated that the total energy consumed during roadway construction is typically on the order of 3 to 7 TJ/lane mile and the total CO_2 emissions during roadway construction varies typically from 200 to 600 Mg/lane mile (1). To make a mile-long, single lane of road in the US, it takes the same energy used by about 50 average American households in one year. As of 2008, the US had almost 8.5 million lane-miles of pavements and 600,000 bridges (18).

During compaction of pavement materials for construction, large quantities of near-potable quality of water are required with significant social and energy impacts. To give an estimate, more than 55 gallons of water per layer per mile are required to compact unbound pavement materials at or near optimum moisture content. Although, most of the water used in road construction is ultimately evaporated from the road and returned to the earth's surface as precipitation in the long run, local depletion can occur rapidly especially in relatively arid areas adding to water scarcity problems (19).

Construction and Maintenance Activities

Potential public safety concerns resulting from roadway construction operations include changing road conditions, lack of consistent or adequate signage, construction emissions, changing pavement surface conditions during construction or rehabilitation, inconsistent or non-existent shoulders, driver distraction due to the presence of construction, night working conditions, etc (14).

Typical issues related to air, water and soil pollution during construction, maintenance and rehabilitation activities include fumes and odors from construction equipment and materials, construction activity emissions, release of pollutants from used construction materials, leachate from construction waste, erosion and sedimentation, heat and energy radiation, etc (14). For instance, pavement sealants are routinely applied to driveways and parking lots as maintenance measures to protect the underlying surfaces. But, these sealants can be a significant source of Polycyclic Aromatic Hydrocarbons (PAHs) which are common pollutants in urban streams adversely affecting the stream biota (20).

Use Phase Impacts

Ageing Highway Infrastructure

Roughly one-third of major roads in the US are in poor or mediocre condition. It costs US motorists $67 billion a year in extra vehicle repairs and operating costs - $324 per motorist by driving on roads in need of repair (see Fig. 4.4). Urban road and highway pavement conditions are likely to worsen at current funding levels (21). It has been reported that a significant investment, much higher than the current level of capital investment, is needed to improve the US highways and roads (22).

Congestion

About 44% of major urban highways are congested costing American motorists $78.2 billion a year in wasted time and fuel costs. The US motorists spend 4.2 billion hours annually stuck in traffic. Due to congestion, roughly 3.9 billion gallons of fuel were wasted in 2009 which is the equivalent of 130 days of flow in the Alaska pipeline (23).

It is estimated that in the US, inadequate capacity of highway infrastructure contributes to 50 percent of congestion (see Fig. 4.5 for a map of potentially congested US highways by 2020) and the remaining 50 percent is caused by crashes, breakdowns, construction work, weather, and special events (24). Roughly, a tenth of the population in the European countries is exposed to what are regarded as dangerous levels of noise from transport, resulting mainly from road traffic (see Fig. 4.6) (25). In addition, there are psychological effects of excessive glare from lighting on roadways and signage. Other social impacts resulting from highway project development may include reduced availability of and accessibility to public space, lack of community cohesion, and greater risk of injury to pedestrians and cyclists (26).

Fig. 4.4. Ageing highway infrastructure costs motorists [*Source:* T. Breen for FHWA (27)]

Vehicle Crashes on Highways

Car crashes is one of the leading causes of death in the US. Over \$230 billion dollars are lost annually due to vehicle crashes on highways which is about 2% of US GDP (28). This amounts to a loss of \$819 for each resident, in medical and insurance costs, lost productivity, travel delay workspace costs, insurance costs and legal costs (29). Almost 42,000 people are killed each year on US highways with an estimated 15,000 traffic fatalities occurring where roadway conditions have been reported to play significant role (28). More than half the driver population perceive that driving is less safe today compared with five years ago (30).

In 2009, 667 workers and motorists were killed and more than 40,000 were injured in highway work zones in the US. Eighty-five percent of those killed in work zones were drivers and their passengers (31).

In assessing the total human health and environmental impacts of vehicle travel, energy and emissions associated with constructing and maintaining the parking infrastructure are usually neglected. According to the International Parking Institute (IPI), an association of parking professionals and the parking industry, there are more than 105 million for-pay US parking spaces (34). When the environmental consequences of parking infrastructure are considered in the automobile's life-cycle, SO_2 and PM_{10} emissions can increase as much as 24% and 89%, respectively, from the baseline inventory (35).

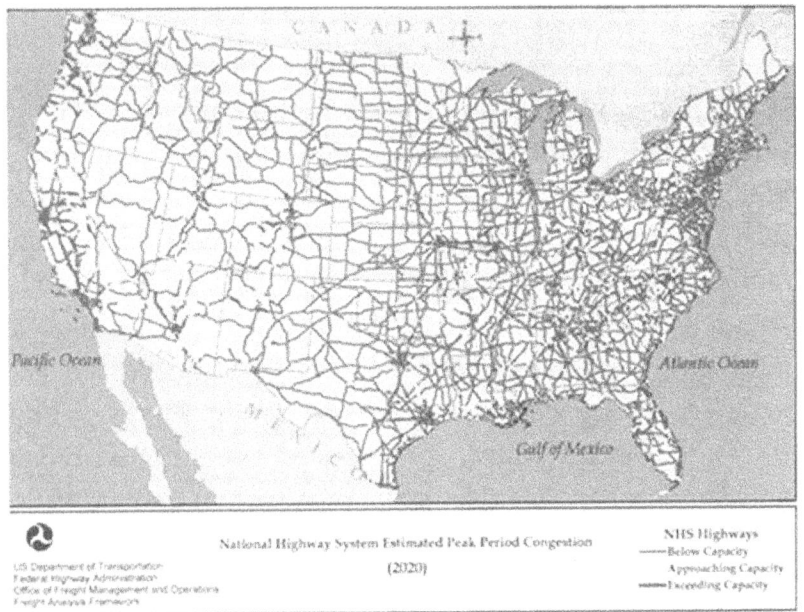

Fig. 4.5. Potentially congested U.S. highways by 2020 [*Source:* FHWA Freight Analysis Framework (32)]

Fig. 4.6. Highway traffic generates dangerous levels of noise and impacts air quality [*Source:* US EPA (33)]

Roadway Deicing

In the US, the use of road salt or sodium chloride (NaCl) for road deicing to improve driving safety in winter dramatically increased in the last 50 years to over 23 million tons in 2005 (36). More than 50% of global salt output is now used by cold countries for deicing roads in winter (see Fig. 4.7). When snowmelt occurs, road salt dissolves in water and the resulting chloride and sodium ions are transported from roads to receiving waters through different pathways (runoff, infiltration, etc.). Finally, they are retained in the surface water and groundwater of a watershed affecting the chemistry and biota in the soil and water (37).

Fig. 4.7. Road salt and other deicing applications during winter have severe ecological impacts [*Source:* US EPA]

Road salt applications have severe environmental consequences especially in the urban areas with high road densities (36):

✓ The water quality in a watershed is degraded by the accumulation of sodium and chloride ions in the environment

✓ Increased chloride concentrations contaminates groundwater and renders water non-potable for human consumption

✓ Increased chloride concentrations decrease the roadside vegetation and biodiversity of wetlands and waterways

✓ Chloride increases the transport and bioavailability of heavy metals in the environment which are also harmful to terrestrial and aquatic biota

A study investigated the road deicing salt impacts on water quality in a major metropolitan area in the US (Twin Cities metropolitan area of Minneapolis/St Paul, Minnesota) and concluded that it would take 10 to 30 years to reach chloride concentrations equal to predevelopment levels if annual inputs of salt to the lakes were stopped now (36).

Pavement Wear

Recent research has shown that dust particles originating from pavement wear are a strong contributor to particulate matter (PM_{10}) concentrations near roads in the Nordic countries during winter and early spring (38). Such non-exhaust emissions include particles from brake and tire wear, road surface abrasion, and particle re-suspension in the wake of passing traffic. In addition, the road surface receive particles from a variety of sources including exhaust particles, deicing salt, grit, biogenic and geogenic material, and hazardous materials, which collectively form an important source of PM emissions (39). With the objective of decreasing air pollution (PM), one of the top priorities in Europe has been to construct road pavements that are more resistant against abrasion.

With the overall goal of achieving reduced fuel consumption and minimizing environmental impacts from tailpipe emissions, policies have been implemented worldwide to reduce freight truck trips by allowing increased loads carried by freight vehicles. However, recent studies have reported that unintended environmental impacts resulting from global warming pollutants (PM_{10}, $PM_{2.5}$, SO_2, CO, Pb, NO_x) may be significant from pavement supply-chain when road freight movement is shifted to heavier vehicles (40).

Urban Heat Island (UHI)

The development of urban areas changes the existing landscape by replacing open land and vegetation by buildings, roads, and other man-made structures. Thus, the sunlight, which would have been absorbed by the vegetation and moist and permeable soil surfaces as part of photosynthesis, is absorbed by the man-made structures with dry and impermeable surfaces. These changes cause an urban area to become significantly warmer than its surrounding rural areas, forming an "island" of higher temperatures in the landscape (see Fig. 4.8). This phenomenon is referred to as Urban Heat Island (UHI) which has serious implications in terms of affecting a community's sustainable environment and quality of life (41).

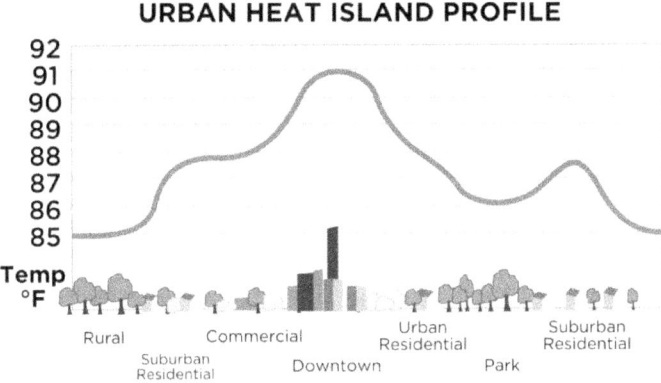

Fig. 4.8. Sketch of an Urban Heat Island (UHI) Profile [*Source:* NASA]

Road pavements constitute about one-quarter the total surface area contributing to the UHI. The negative UHI impacts include increased energy consumption, elevated GHG emissions, compromised human health and comfort, and impaired water quality. UHI effect also leads to a greater use of air conditioning which further increases the ozone concentration and air pollution (42). Fig. 4.9 displays satellite images of Atlanta to illustrate how highways and buildings contribute to UHI.

Environmental Life Cycle Assessment (LCA) Framework

To properly assess and appropriately mitigate environmental impacts from pavements, it is necessary for decision makers to consider their life-cycle energy consumption, emissions, and wastes. The Life Cycle Assessment (LCA) is a tool for characterizing and quantifying environmental sustainability of a product/process or system throughout its entire life cycle (44,45).

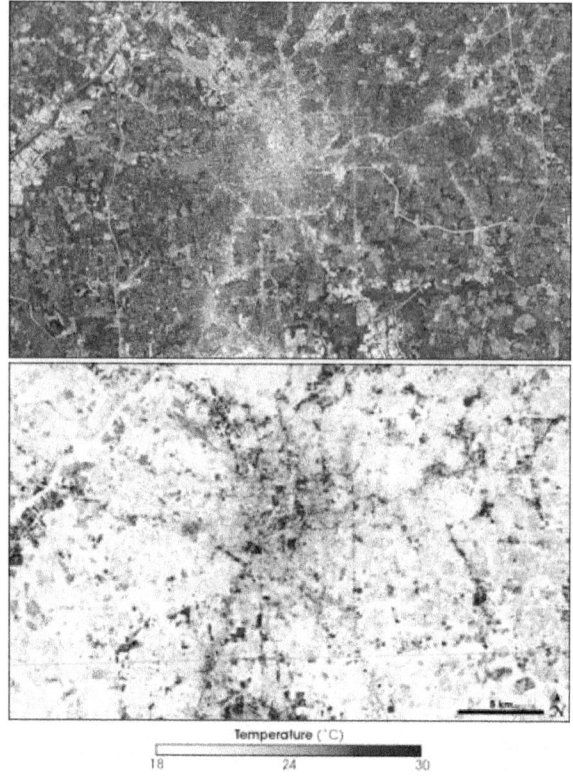

Temperature (°C)
18   24   30

Fig. 4.9. Photographic (top) and infrared view (below) of Atlanta from satellite images showing how highways and buildings contribute to urban heat island [*Source:* NASA (43)]

Using a "cradle-to-grave" perspective, environmental and sustainability impacts associated with all stages of a process or product, from raw materials through materials processing, manufacturing, use, repair and maintenance, and disposal or recycling, are considered in LCA (46). Fig. 4.10 depicts the major life cycle stages that can be considered in an LCA and the typical inputs/outputs measured (46). Note that LCA is different from the Life Cycle Cost Analysis (LCCA) which evaluates the *total cost* of a project or product over its service life.

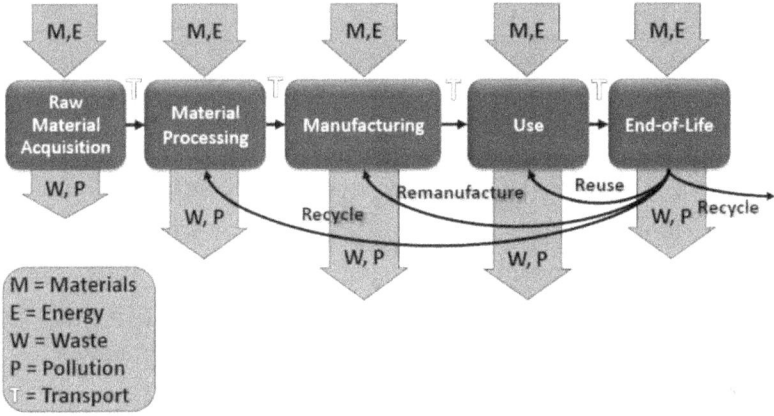

Fig. 4.10. Major life cycle stages in product or process development [*Source*: G. A. Keolein and A. M. Kendall]

Common categories of social and environmental impacts and damages attributable to the life cycle of a product include (46):

✓ Global Warming Potential (GWP) of GHG emissions

✓ Acidification (soil and ocean)

✓ Smog

✓ Ozone layer depletion

✓ Eutrophication

✓ Human health

✓ Terrestrial toxicity

✓ Aquatic toxicity

✓ Noise

✓ Habitat destruction

✓ Desertification

✓ Land and water use

✓ Depletion of minerals and fossil fuels

LCA provides a more accurate picture of the true environmental trade-offs in product or process selection by compiling an inventory of relevant energy and material inputs and environmental releases and evaluating their potential environmental impacts and finally interpreting the results to make a more informed decision (46). Since the inception of its concept in the 1960s, LCA has evolved significantly and was standardized by the International Standards Organization (ISO) in the late 1990s (47), to establish a uniform standard for making impact assessments and to prevent manufactures from using it inappropriately to make broad marketing claims (46).

According to the ISO's definition (47), LCA is a process that "addresses the environmental aspects and potential environmental impacts (e.g. use of resources and the environmental consequences of releases) throughout a product's lifecycle from raw material acquisition, through production, use, end-of-life treatment, recycling and final disposal (i.e. cradle-to-grave)." A generalized framework for many attributes and issues that are relevant to LCA is provided in the ISO 14040 series of standards although the mechanics of conducting an LCA are not discussed (47). A LCA is carried out in four phases as shown in Fig. 4.11 (46,47):

• *Goal Definition and Scoping*: This step defines the life-cycle system boundaries, functional unit used for the analysis, and the intended audience. This will outline how the study will be conducted and what are the expected final results. The system boundaries determine the unit processes and associated emissions that will be included in the LCA. In relation to the intended application, the functional unit is defined as the "quantified performance of a product system for use as a reference unit". (47) It provides a basis for comparing and analyzing alternative products or services.

- *Lifecycle Inventory Analysis (LCI):* This phase identifies and quantifies the consumption of resources (energy, water and materials), waste flows, and environmental releases (e.g., air emissions, solid waste disposal, waste water discharges) from all processes within the life-cycle system boundaries. Apart from data collection and processing, and calculations, the inventory analysis also includes allocation of environmental flows and releases, and other materials or byproducts that were specified in the previous phase.

- *Lifecycle Impact Assessment (LCIA):* Impact assessment organizes resource inputs and emissions from inventory analysis into environmental impact categories (e.g., global warming potential, acidification, and primary energy use) for more meaningful assessment of the life-cycle inventory results.

- *Interpretation:* The results of the inventory analysis and impact assessment are evaluated to draw conclusions, select the preferred product or process with a clear understanding of the uncertainty and assumptions used to generate the results, and make recommendations to aid in the decision-making process. Decision-making based on LCI and LCIA results can further be assisted by adopting an existing decision-making weighting scheme, using the Analytical Hierarchy Process (AHP), Modified Delphi Technique, or a Multi-Attribute Utility Theory (MAUT) (18).

LCA is widely used in a number of applications as it has several sustainability benefits (46). LCA can:

✓ Help identify transfer of environmental impacts from one media to another and/or from one life cycle stage to another

✓ Analyze environmental trade-offs with one or more products/processes

✓ Promote awareness of production efforts and link them to system consumption

✓ Use of comprehensive social, economic, and environmental accounting metrics to measure performance

✓ Help select product or process with least impacts on the environment

✓ Identify opportunities to improve environmental performance at different points in the life-cycle

✓ Help evaluate relevance of various indicators for environmental performance

✓ Provide marketing opportunity such as eco-labeling

✓ Compare health and ecological impacts between two or more rival products/processes

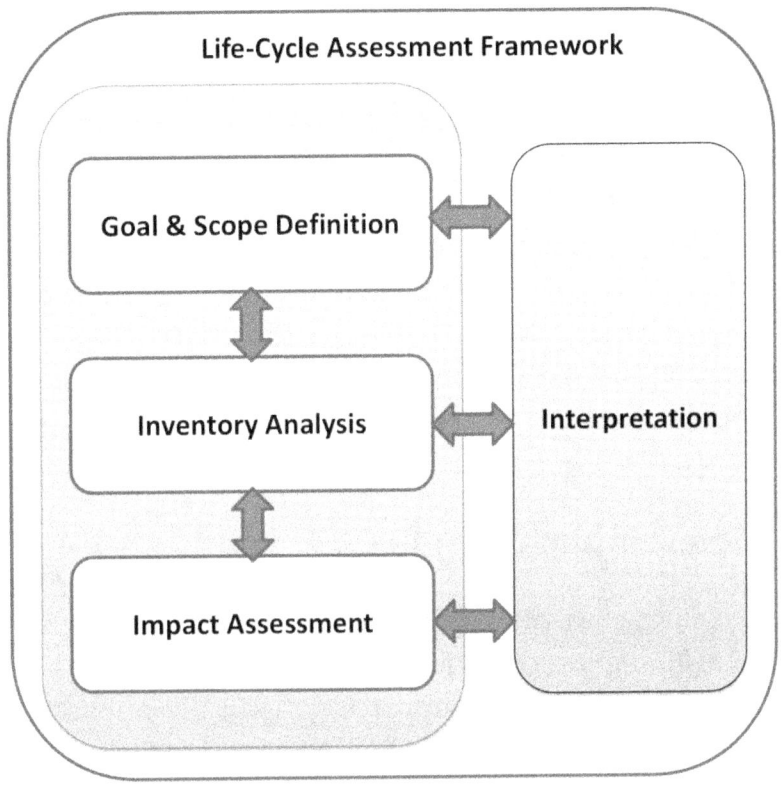

Fig. 4.11. Four phases of LCA [*Source:* ISO (47)]

LCA Approaches

The three different approaches to conducting an LCA include: (1) Process-based LCA; (2) Economic Input-Output LCA (EIO-LCA); and (3) Hybrid EIO-LCA (a combination of process LCA and IO-LCA, also called hybrid EIO-LCA).

Process-based LCA

The ISO 14040 Standards (47) are primarily concerned with process-based LCA, which is focused on analyzing the actual process (i.e., acquisition of raw materials, manufacturing, transportation, etc.) involved in making a product. The BEES (Building for Environmental and Economic Sustainability) LCA software developed by National Institute of Standards and Technology (NIST) is a good example of a process-based LCA tool.

Economic Input-Output LCA

Most countries in the world routinely produce such Input-Output (IO) tables for various industry sectors. For instance, the US Department of Commerce periodically publishes IO tables for the US economy. EIO-LCA methods aggregate such IO tables with sector-level environmental data to quantify economy-wide environmental impacts associated with producing a given product or service (50).

EIO-LCA overcomes the problems associated with conventional process-based LCA which captures only those emissions that occur within the semi-arbitrarily-drawn system boundary and does not account for indirect effects. IO-LCA addresses these problems by using the IO tables to trace the flow of products or services through the economy. Researchers often use IO-LCA as a screening tool to estimate the relative importance of indirect resource use or emissions which can be used to guide a process-based LCA (51).

Hybrid EIO-LCA

Hybrid LCA uses IO-LCA to fill the gaps left by process-based LCA to produce a more comprehensive representation of the environmental effects of the system processes (52).

There are different ways in which a hybrid LCA can be implemented depending on individual project constraints and data availability. There are four primary categories of hybrid LCA reported in the literature: tiered hybrid analysis, IO bases hybrid analysis, integrated hybrid analysis, and augmented process-based hybrid analysis (53).

Pavement LCA

Pavement LCA is a relatively new area of study that is gaining lot of attention in recent years owing to its usefulness in assessing the environmental impacts of pavements. A comprehensive review of existing literature and research on LCA of pavements was recently completed by researchers from University of California (Berkeley) and Lawrence Berkeley National Laboratory (LBNL) to determine the current state of practice, establish a baseline for the current state of research, and identify knowledge and implementation gaps (48).

Another latest development in this regard is the Pavement LCA Workshop (54) conducted in May 2010 by the University of California (UC) Pavement Research Center (UCPRC, Davis and Berkeley) and the University of California Institute of Transportation Studies (Berkeley and Davis), who are working together with the International Society for Asphalt Pavements (ISAP) and the International

Society for Concrete Pavements (ISCP) on establishing common practices for conducting environmental LCA for pavements. As part of the workshop, a draft document was prepared by the Pavement LCA Group at UC Davis to provide preliminary system definitions and standard assumptions for basic elements of pavement LCA (55). The Office of Asset Management, Pavements and Construction at the FHWA have developed an entire website dedicated to sustainable   pavements: https://www.fhwa.dot.gov/pavement/sustainability/. Most recently, a comprehensive edited book has published by Springer-Verlag on the topic of sustainable pavements: http://www.springer.com/us/book/9783662447185.

Software Tools for Pavement LCA

Among the existing software tools designed for assessing the pavement life cycle and used by many State DOTs, a vast majority of them are concerned only with the Life-Cycle Cost Analysis (LCCA) such as the FHWA's RealCost (56).

LCA models that are specifically oriented towards assessing the environmental impacts of pavements are less than a handful: PaLATE (Pavement Life-Cycle Assessment Tool for Environmental and Economic Effects), the Danish ROAD-RES software, BenReMod-LCA (Beneficial Reuse Modules), and IRF CHANGER (the Calculator for Harmonised Assessment and Normalisation of Greenhouse-gas Emissions for Roads).

PaLATE

The PaLATE (Pavement Life-Cycle Assessment Tool for Environmental and Economic Effects) LCA tool (57,58) was developed in 2003 by the consortium of Green Design at the University of California, Berkeley to estimate the LCA of environmental and economic effects associated with pavements and roads. It is considered a Streamlined LCA (SCLA) or short-hand version of the full-scale LCA which models the material extraction, production, construction, maintenance, and end-of-life phases of the pavement life-cycle.

ROAD-RES

The Danish ROAD-RES model (5) is a LCA tool for road construction and disposal of residues. As an environmental management decision support tool for the road sector, ROAD-RES can assist in designing new roads as well as the operation and maintenance of existing roads. It can also be used as a decision support tool to manage sustainable disposal of residues, either by landfilling or utilization in roads. It can be used to assess leaching from all road construction materials, including granular and bound materials, and landfilled granular materials (5).

BenReMod-LCA

BenReMod (Beneficial Reuse Modules)-LCA (59) is a web-based LCA tool (under development) for comparing environmental impacts of virgin materials and industrial by-products (such as coal fly ash, coal bottom ash, and recycled concrete pavement) in road construction. It employs a process-based LCA approach. Construction activities such as excavation and compaction as well as maintenance are not considered, since it is assumed that environmental emissions and cost associated would be similar for maintenance work for both natural aggregates and recycled materials (59).

The environmental impact categories used by BenReMod-LCA include: (i) cost, (ii) energy consumption, (iii) acidification potential, (iv) global warming potential (GWP), (v) human toxicity potential (HTP), (vi) aquatic ecotoxicity potential (FAETP), (vii) aquatic sediment ecotoxicity potential (FSETP), and (viii) terrestrial ecotoxicity potential (TETP) (59).

IRF CHANGER

The International Road Federation (IRF) has designed a GHG calculator, CHANGER (the Calculator for Harmonised Assessment and Normalisation of Greenhouse-gas Emissions for Roads), specifically tailored to road infrastructure projects based on Input-Output (IO) modeling approach (60). The system contains data from 188 countries covering all global regions and income groupings and is fully compatible with the IPCC recommendations.

The IRF GHG calculator is divided into four main modules: pre-construction, pavement, security barriers, and road signs. The pre-construction module takes into account clearing and piling, cut exports and fill imports transport to and from the road site. The pavement module considers on-site impacts, pavement construction materials, materials transport, and construction machines. Work is currently being carried out to incorporate a new module devoted to maintenance activities (60).

The input data entered by the user of the calculator is used to obtain the material quantities, material transport, electricity used, etc., which are finally compiled with emissions factors to calculate the total emissions related to every step of the road construction process. The LAVOC (Traffic Facilities Laboratory) of the Swiss Federal Institute of Technology (Ecole Polytechnique Fédérale de Lausanne - EPFL) has analyzed and validated both the quality and reliability of the CHANGER's databases and the calculation procedures. It is noted that CHANGER is not intended to be a complete LCA and the GHG emissions considered are limited and some important emissions to air, water or land are not taken into account (61).

Other Pavement LCA Tools and GHG Calculators

Other sustainability impact assessment tools for highways and pavements (under development) based on LCA methodology include:

- ✓ *EESAR* (Environmental, Economic and Social Assessment of Roads) – Finland (62)

✓ *ERM/GRM* (Elementary Road Modulus/Global Road Modulus) - The Sustainable Development Team of Laboratoire Central des Ponts et Chaussées (LCPC) of France (63)

✓ *GreenDOT* (64): A spreadsheet-based calculator tool that calculates CO_2 emissions from the operations, construction, and maintenance activities of US state DOTs

✓ *National Asphalt Pavement Association's (NAPA's) GHG Calculator* (65): An online calculator to help asphalt mix facility engineers understand the amount of GHGs their facilities release

✓ *American Concrete Pavement Association (ACPA's) GreenStreet Calculator* (66): An online calculator to help one see the positive contribution concrete pavements can make as part of a "Green Streets" program

Useful Online Resources

US EPA's LCA Resources including Software and Databases

http://www.epa.gov/nrmrl/lcaccess/resources.html

PaLATE

http://www.ce.berkeley.edu/~horvath/palate.html

Economic Input-Output LCA (EIO-LCA)

http://www.eiolca.net/

FHWA Life-Cycle Cost Analysis (LCCA)

http://www.fhwa.dot.gov/infrastructure/asstmgmt/lcca.cfm

Pavement LCA Workshop (May 5-7, 2010)

http://www.ucprc.ucdavis.edu/p-lca/index.html

International Sustainable Pavements Workshop (Jan. 7-9, 2010)

http://www.vtti.vt.edu/ISPW/ISPW-2010.html

International Road Federation (IRF) GHG Calculator

http://www.irfghg.org/

NAPA's GHG Calculator

http://hotmix.org/ghgc/ghgcv3.html

ACPA's GreenStreet Calculator

http://1734298.sites.myregisteredsite.com/green11/calculator.aspx

MIT Concrete Sustainability Hub

http://web.mit.edu/cshub/

References

1.   Muench ST. Roadway Construction Sustainability Impacts. Transportation Research Record: Journal of the

Transportation Research Board [Internet]. 2010 [cited 2011 Mar 16];2151:36-45. Available from: http://trid.trb.org/view.aspx?id=910627

2. Rodrigue J-P, Comtois C, Slack B. The geography of transport systems [Internet]. Taylor & Francis; 2009 [cited 2011 Mar 17]. 369 p.Available from: http://people.hofstra.edu/geotrans/eng/ch8en/conc8en/ch8c1en.html

3. Mallick RB, Veeraraghavan A. Sustainable Pavements in India: The Time To Start Is Now. New Building Materials and Construction World (NBM&CW) Magazine [Internet]. 2010 Sep [cited 2011 Feb 9];16(3). Available from: http://www.nbmcw.com/articles/roads/18233-sustainable-pavements-in-india-the-time-to-start-is-now.html

4. USGS. Mineral Commodity Summaries 2010. Washington, D.C.: U.S. Geological Survey (USGS); 2010.

5. Birgisdóttir H. Life Cycle Assessment Model for Road Construction and Use of Residues from Waste Incineration [Internet] [Ph.D. Dissertation]. [Kgs. Lyngby, Denmark]: Technical University of Denmark; 2005 [cited 2011 Feb 8]. 60 p.Available from: http://orbit.dtu.dk/getResource?recordId=181724&objectId=1&versionId=1

6. USGS. USGS Fact Sheet FS 144-97: Natural Aggregates—Foundation of America's Future [Internet]. 1999 Feb [cited 2011 Mar 17];Available from: http://minerals.usgs.gov/minerals/pubs/commodity/aggregates/fs14497.pdf

7.   Zapata P, Gambatese JA. Energy consumption of asphalt and reinforced concrete pavement materials and construction. Journal of infrastructure systems. 2005;11(1):9-20.

8.   Zhang H. Sustainable Pavement Asset Management Based on Life CycleModels and Optimization Methods [Internet] [Ph.D. Dissertation]. [Ann Arbor, MI]: University of Michigan; 2009 [cited 2011 Mar 16]. 147 p.Available from: http://css.snre.umich.edu/css_doc/CSS09-14.pdf

9.   Wilson A. Cement and Concrete: Environmental Considerations. Environmental Building News [Internet]. 1993 Mar [cited 2011 Mar 17];2(2). Available from: http://www.buildinggreen.com/auth/article.cfm/1993/3/1/Cement-and-Concrete-Environmental-Considerations/

10. Zimmermann M, Althaus H-J, Haas A. Benchmarks for sustainable construction: A contribution to develop a standard. Energy and Buildings [Internet]. 2005 Nov [cited 2011 Feb 9];37(11):1147-1157. Available from: http://www.sciencedirect.com/science/article/B6V2V-4GX0CBC-8/2/6a57cf5bb3d32cbdf4f3bd0cfad5806f

11. Gerilla GP, Takeyama Y, Inamura H. Environmental Impacts of the Road and Building Construction Industry in Japan [Internet]. In: Proceedings of the International Symposium on Lowland Technology. Saga, Japan: 2000 [cited 2011 Feb 9]. p. 17.Available from: http://people.revoledu.com/gloria/pdf/ISLT2000.pdf

12. Calkins M. Materials for sustainable sites: a complete guide to the evaluation, selection, and use of sustainable construction materials. John Wiley and Sons; 2009. 466 p.

13. Saride S, Balunaini U. Towards Green Pavements in India [Internet]. In: Proceedings of The First US-India Workshop on Global Geoenvironmental Engineering Challenges. New Delhi, India: 2010 [cited 2011 Feb 8]. p. 6.Available from: http://tigger.uic.edu/~kreddy/GGEC/site_includes/Siree sh-Whitepaper.pdf

14. PIARC. Impact of Road Construction and Maintenance Activities on Road Users and the Adjacent Land Use [Internet]. France: World Road Association (PIARC); 2007 [cited 2011 Mar 27]. Available from: http://publications.piarc.org/ressources/publications_fil es/4/1934,2007R05-WEB.pdf

15. FHWA. Chapter 1: Page 2 - Freeway Management and Operations Handbook [Internet]. 2009 May 22 [cited 2011 Mar 17];Available from: http://ops.fhwa.dot.gov/freewaymgmt/publications/frw y_mgmt_handbook/chapter1_02.htm

16. Damrad T, Bekker GJ. COST 341 - Habitat Fragmentation Due to Transportation Infrastructure: Findings of the COST Action 341 [Internet]. Luxembourg, Europe: Office for Official Publications of the Eurpean Communities; [cited 2011 Feb 10]. Available from: http://www.iene.info/cost-341/finalreportdef.pdf

17. FHWA. Designing the Next Generation ofWildlife Crossings. Focus [Internet]. 2011 Mar [cited 2011 Mar

29];:8. Available from:
http://www.fhwa.dot.gov/publications/focus/11mar/11
mar.pdf

18. Muench ST, Anderson JL, Hatfield JP, Koester JR,
Soderlund M. Greenroads Rating System v1.5. Seattle,
WA: University of Washington;

19. Steyn W, Paige-Green P. Evaluation of Issues Around
Road Materials for Sustainable Transport [Internet]. In:
Proccedings of the 28th Annual Southern African
Transport Conference (SATC). Pretoria, South Africa:
Southern African Transport Conference; 2009 [cited
2011 Feb 8]. p. 11.Available from:
http://researchspace.csir.co.za/dspace/handle/10204/370
0

20. US EPA. Water/Sediment Quality - Pavement Sealants
[Internet]. CADDIS: Sources, Stressors & Responses.
2010 Sep 20 [cited 2011 Mar 18];Available from:
http://www.epa.gov/caddis/ssr_urb_wsq4.html

21. TRIP. Hold the Wheel Steady:America's Roughest Rides
and Strategies to Make our Roads Smoother [Internet].
Washington D.C.: TRIP: A National Transportation
Research Group; 2010 [cited 2011 Feb 10]. Available
from:
http://www.tripnet.org/Urban_Roads_Report_Sep_2010
.pdf

22. ASCE. ASCE 2009 Infrastructure Fact Sheet: Facts
About Roads [Internet]. 2009 [cited 2011 Feb 10].
Available from:
http://www.infrastructurereportcard.org/sites/default/fi
les/RC2009_roads.pdf

23. Texas Transportation Institute (TTi). 2010 Annual
    Urban Mobility Report: Urban Mobility Information
    [Internet]. Press Release: Media Information. 2011 Jan 20
    [cited 2011 Feb 10];Available from:
    http://mobility.tamu.edu/ums/media_information/press
    _release.stm

24. AASHTO. Transportation Invest In Our Future: A New
    Vision for the 21st Century [Internet]. Washington,
    D.C.: American Association of State and Highway
    Transportation Officials (AASHTO); 2007 [cited 2011
    Mar 16]. Available from:
    http://www.transportation1.org/tif5report/TIF5.pdf

25. Gilbert R, Irwin N, Hollingworth B, Blais P, et al.
    Sustainable Transportation Performance Indicators
    (STPI) Project: Report on Phase 3 of the STPI Project
    [Internet]. Canada: The Centre for Sustainable
    Transportation (CST); 2002 [cited 2011 Mar 17].
    Available from:
    http://cst.uwinnipeg.ca/documents/STPI%20Phase%203
    %20report.pdf

26. Forkenbrock DJ, Weisbrod GE. Assessing the Social
    andEconomic Effects ofTransportation Projects
    [Internet]. Washington, D.C.: Transportaion Research
    Board (TRB); 2001 [cited 2011 Mar 25]. Available from:
    http://onlinepubs.trb.org/onlinepubs/nchrp/nchrp_w31.
    pdf

27. Taylor R. Solutions To the Traffic Snarl. Public Roads
    [Internet]. 2007 [cited 2011 Mar 17];71(1). Available
    from:
    http://www.fhwa.dot.gov/publications/publicroads/07ju
    ly/05.cfm

28. Donahue TJ. Statement of the U.S. Chamber of Commerce on Examining the Stimulative Effects of Infrastructure Investment on the U.S. Economy and the Need for Additional Investment [Internet]. 2009 Jan 22 [cited 2011 Feb 10];Available from: http://www.fasterbettersafer.org/images/stories/pdfs/tjd%20ti%20testimony%2001-09.pdf

29. TRIP. Key Facts About America's Surface Transportation System and Federal Funding [Internet]. 2011 Jan [cited 2011 Feb 10];Available from: http://www.tripnet.org/Fact_Sheet_National.pdf

30. AAA. 2010Traffic SafetyCulture Index [Internet]. Washington D.C.: AAA Foundation for Traffic Safety; 2010 [cited 2011 Feb 10]. Available from: http://www.aaafoundation.org/pdf/2010TSCIndexFinalReport.pdf

31. FHWA. Safer Driving, Safer Work Zones:National Work Zone Awareness Week 2011. Focus [Internet]. 2011 Mar [cited 2011 Mar 29];:8. Available from: http://www.fhwa.dot.gov/publications/focus/11mar/11mar.pdf

32. FHWA. An Initial Assessment of Freight Bottlenecks on Highways - 2.0 [Internet]. 2011 Feb 14 [cited 2011 Mar 17];Available from: http://www.fhwa.dot.gov/policy/otps/bottlenecks/chap2.htm

33. US EPA. Our Nation's Air: Status and Trends Through 2008 [Internet]. Washington, D.C.: US Environmental Protection Agency (EPA); 2010 [cited 2011 Mar 18]. Available from: http://www.epa.gov/airtrends/2010/

34. IPI. International Parking Institute (IPI) [Internet]. 2009 [cited 2011 Feb 9];Available from: http://www.parking.org/home-page.aspx

35. Chester M, Horvath A, Madanat S. Parking infrastructure: energy, emissions, and automobile life-cycle environmental accounting. Environ. Res. Lett. [Internet]. 2010 [cited 2011 Feb 9];5(3):8. Available from: http://iopscience.iop.org/1748-9326/5/3/034001

36. Novotny EV. Road Deicing Salt Impacts on Urban Water Quality [Internet] [Ph.D. Thesis]. [Twin Cities, Minnesota]: University of Minnesota; 2009 [cited 2011 Feb 8]. 205 p.Available from: http://home.safl.umn.edu/bmackay/1Pub/Theses/Novotny_Eric_PhD_2009.pdf

37. Thunqvist E-L. Regional Increase of Mean Chloride Concentration in Water Due to the Application of Deicing Salt. Science of The Total Environment [Internet]. 2004 [cited 2011 Feb 9];325(1-3):29-37. Available from: http://www.sciencedirect.com/science/article/B6V78-4BYJYXX-4/2/d7dea1b7a250e02e67a7586c6471351f

38. Kupiainen K, Tervahattu H, Räisänen M. Experimental studies about the impact of traction sand on urban road dust composition. The Science of the Total Environment. 2003;308(1-3):175–184.

39. Praticò FG, Ammendola R, Moro A. Factors affecting the environmental impact of pavement wear. Transportation Research Part D: Transport and Environment [Internet]. 2010 May [cited 2011 Feb 9];15(3):127-133. Available from:

http://www.sciencedirect.com/science/article/B6VH8-
4YGHKJR-1/2/3c94cd30f3a060585f346e3bb57cf261

40. Sathaye N, Horvath A, Madanat S. Unintended impacts
    of increased truck loads on pavement supply-chain
    emissions. Transportation Research Part A: Policy and
    Practice [Internet]. 2010 Jan [cited 2011 Feb 9];44(1):1-15.
    Available from:
    http://www.sciencedirect.com/science/article/B6VG7-
    4XJP5T2-1/2/904af24cb77258752c2ae645cc6249e4

41. Rose LS, Akbari H, Taha H. Characterizing the fabric of
    the urban environment: A case study of Greater
    Houston, Texas [Internet]. Berkeley, CA: Lawrence
    Berkeley National Laboratory (LBNL); 2003 [cited 2011
    Feb 23]. Available from:
    http://www.osti.gov/bridge/servlets/purl/816533-
    RZAzAi/native/

42. Akbari H, Menon S, Rosenfeld A. Global cooling:
    increasing world-wide urban albedos to offset CO_2.
    Climatic Change [Internet]. 2008 [cited 2011 Feb
    13];94(3-4):275-286. Available from:
    http://www.springerlink.com/content/r465853147015k4
    g/

43. NASA. Measuring Temperature Islands [Internet]. My
    NASA Data: Mentoring and Inquiry Using NASA Data
    on Atmospheric and Earth Science for Teachers and
    Amateurs. 2011 Jan 6 [cited 2011 Mar 18];Available
    from: http://mynasadata.larc.nasa.gov/P16.html

44. Ciambrone DF. Environmental life cycle analysis. CRC
    Press; 1997. 162 p.

45. Curran MA. Environmental life-cycle assessment.
    McGraw-Hill; 1996. 432 p.

46. US EPA. Life Cycle Assessment: Principles and Practice [Internet]. Cincinnati, Ohio: National Risk Management Research Laboratory, US EPA; 2006 [cited 2011 Feb 10]. Available from: http://www.epa.gov/nrmrl/lcaccess/pdfs/600r06060.pdf

47. International Organization for Standardization. Environmental management -- Life cycle assessment -- Principles and framework. Geneva, Switzerland: International Organization for Standardization; 2006.

48. Santero N, Masanet E, Horvath A. Life Cycle Assessment of Pavements:A Critical Review of Existing Literature and Research [Internet]. Berkeley, CA: Lawrence Berkeley National Laboratory (LBNL); 2010 [cited 2011 Feb 10]. Available from: www.osti.gov/bridge/servlets/purl/985846-cyLjIR/

49. Keoleian GA, Meneray D. Life Cycle Design Manual: Environmental Requirements and The Product System. Washington D.C.: Office of Research and Development, US EPA; 1993.

50. Hendrickson CT, Lave LB, Matthews HS. Environmental life cycle assessment of goods and services: an input-output approach. Washington D.C.: Resources for the Future; 2006.

51. Suh S, editor. Handbook on Input-Output Economics in Industrial Ecology [Internet]. Springer Science+Business Media B.V.; 2009 [cited 2011 Feb 11]. 885 p.Available from: http://www.springer.com/new+%26+forthcoming+titles+%28default%29/book/978-1-4020-4083-2

52. Treloar GJ, Love PED, Faniran OO, Iyer-Raniga U. A hybrid life cycle assessment method for construction. Construction Management and Economics. 2000;18(1):5–9.

53. Suh S. Functions, commodities and environmental impacts in an ecological-economic model. Ecological Economics [Internet]. 2004 [cited 2011 Feb 11];48(4):451–467. Available from: http://linkinghub.elsevier.com/retrieve/pii/S0921800904000229

54. UCPRC P-LCA Team. Pavement Life Cycle Assessment Workshop [Internet]. 2010 Sep [cited 2011 Feb 14];Available from: http://www.ucprc.ucdavis.edu/p-lca/

55. Harvey J, Santero N, with Kendall A, Lee and Wang. Pavement Life Cycle Assessment: System Definition and Standard Assumptions [Internet]. Davis, California: University of California; 2010 [cited 2011 Feb 10]. Available from: http://www.ucprc.ucdavis.edu/p-lca/pdf/plca_document_draft_web.pdf

56. FHWA. FHWA Life-Cycle Cost Analysis (LCCA) Software - RealCost Version 2.5 [Internet]. 2011 Apr 5 [cited 2011 Apr 5];Available from: http://www.fhwa.dot.gov/infrastructure/asstmgmt/lccasoft.cfm

57. Horvath A. PaLATE - Program and User Guide - Version 2.0 [Internet]. Recycled Materials Resource Center (RMRC). 2004 Apr [cited 2011 Mar 13];Available from: http://www.rmrc.unh.edu/Resources/CD/PaLATE/PaLATE.htm

58. Horvath A. PaLATE - Pavement Life-cycle Assessment Tool for Environmental and Economic Effects [Internet]. Consotrium on Green Design and Manufacturing: University of California, Berkeley. 2007 Jun 5 [cited 2011 Mar 13];Available from: http://www.ce.berkeley.edu/~horvath/palate.html

59. Chowdhury R, Apul D, Fry T. A life cycle based environmental impacts assessment of construction materials used in road construction. Resources, Conservation and Recycling [Internet]. 2010 Feb [cited 2011 Feb 15];54(4):250-255. Available from: http://www.sciencedirect.com/science/article/B6VDX-4XDFDMN-1/2/4fc47935fac6315b5c37a582d3521d72

60. IRF. CHANGER - Green House Gas Calculator - International Road Federation [Internet]. 2010 [cited 2011 Feb 15];Available from: http://www.irfghg.org/

61. Bueche N, Dumont A-G. IRF Greenhouse Gas Calculator - Analysis and Validation - Excerpt from EPFL Report. Lausanne, Switzerland: École Polytechnique Fédérale de Lausanne (EPFL); 2009.

62. Sirvio K, Sari J. Software Tool and Model for Economic, Environmental and Social Impact Assessment of Road Projects [Internet]. In: Proceedings of the 12th World Congress on Transportation Research (WCTR). Lisbon, Portugal: World Conference on Transport Research Society (WCTRS); 2010 [cited 2011 Mar 27]. p. 13.Available from: http://www.civil.ist.utl.pt/wctr12_lisboa/WCTR_General/documents/02165.pdf

63. Hoang T, Jullien A, Ventura A. A global methodology for sustainable road -Application to the environmental

assessment of French highway [Internet]. In: Proceedings of the 10th International Conference On Durability of Building Materials and Components (DBMC). Lyon, France: 2005 [cited 2011 Mar 16]. p. 10.Available from: http://www.irbdirekt.de/daten/iconda/06089002852.pdf

64.   Gallivan F. Greenhouse Gas Mitigation Measures for Transportation Construction,maintenance, and Operations Activities [Internet]. Washington, D.C.: Transportaion Research Board (TRB); 2010 [cited 2011 Apr 4]. Available from: http://onlinepubs.trb.org/onlinepubs/nchrp/docs/NCHRP25-25%2858%29_FR.pdf

65.   NAPA. Asphalt Mix Plant Greenhouse Gas Calculator: Total Facility Emissions [Internet]. National Asphalt Pavement Association (NAPA). [cited 2011 Apr 4];Available from: http://hotmix.org/ghgc/ghgcv3.html

66.   ACPA. Green Streets Calculator [Internet]. American Concrete Pavement Association (ACPA). 2009 [cited 2011 Apr 4];Available from: ttp://1734298.sites.myregisteredsite.com/green11/calculator.aspx

Chapter 5:
Recycled and Reused Materials in Highways

In the US, roadway and highway construction currently consumes between 500 and 800 million metric tons of virgin crushed rock, gravel, and sand each year as aggregates used in road construction (1). As we have already seen, the social, economic, and environmental impacts associated with the extraction of raw materials and transportation of these materials between facilities for pavement construction and rehabilitation is quite significant in the overall life cycle of a pavement.

There is an urgent need to significantly reduce the amount of virgin materials required for roadway construction, address energy and climate change issues arising from extraction of raw materials, and reduce the need for landfill space for pavements that are disposed of (1). In this regard, the well-known "3R" concept (reduce, reuse, and recycle) and the waste hierarchy principles (see Fig. 5.1), which encourage the effective use of resources by balancing environmental conservation and economic growth is very much applicable to pavements (2). In addition, the recovered materials should be used in such a way that the expected pavement performance will not be compromised.

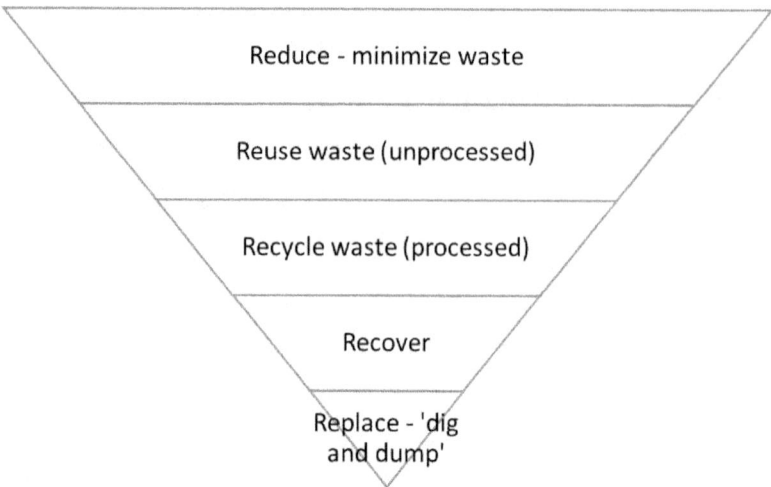

Fig. 5.1. The waste hierarchy [*Source:* UK Roads Board (3)]

According to the *FHWA's Recycled Materials Policy* (4) issued on February 7, 2002, recycling of aggregates and other highway construction materials from old, deteriorated pavements makes a sound economic, environmental, and engineering sense, and can be a cost-saving measure, freeing funds for additional highway construction, rehabilitation, preservation or maintenance.

A shift in thinking is required to consider the existing roadway as a giant linear quarry that is in all areas, unlike the natural quarries, and a source of "technical nutrients" for the reconstruction of new pavement and an opportunity to eliminate waste (5,6). The pavement sustainability framework (see Fig. 5.2) should emphasize that pavements are constructed for future reuse/recycling (7). The future reuse of the same pavement or how it can be recycled into a new pavement or use should be considered. Even those materials which don't enter the closed cycle of pavement use and maintenance can be utilized in other applications.

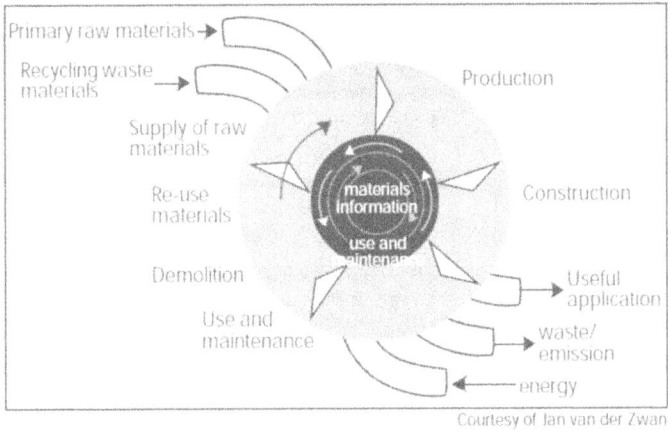

Courtesy of Jan van der Zwan

Fig. 5.2. Sustainable recycling strategy cycle [*Source:* FHWA (7)]

Highway agencies worldwide have established standard specifications regulating the reuse and recycling of highway construction materials as well as beneficial reuse of wastes and by-products from other industries into construction of new pavements (7). Most often, the incorporation of such innovative recycled materials has led to an improvement in pavement performance. An well-known example is the partial replacement of Portland cement with fly ash (a by-product from coal-fired power plants) in concrete which not only enhances the durability, but also saves the energy and greenhouse gas emissions associated with producing cement (8).

Recognizing that sustainable choices about materials for highway construction and maintenance is not something that should be done in isolation or on an opportunistic basis, the UK Roads Board encourages adopting a multi-step participatory strategy for incorporating sustainability into highway construction and maintenance, especially in relation to the use of recycled materials (see Fig. 5.3) (3).

- Get Support from the top
- Get all the key players involved at an early stage
- Appoint a champion
- Assess what is appropriate for your area
- Liase with other departments in your council
- Ensure you have logistic support for recycling
- Develop a procurement strategy to encourage sustainability
- Do not be afraid to use primary aggregates if appropriate
- Set appropriate key performance indicators
- Insist on quality control from suppliers
- Keep track of materials
- Create a sustainability culture throughout the supply chain
- Monitor progress
- Aim for continual improvement
- Celebrate success

Fig. 5.3. Steps recommended by UK Roads Board for incorporating sustainability into highway maintenance and construction [*Source:* UK Roads Board (3)]

Reclaimed Asphalt Pavement (RAP)

It is estimated that about 100 million tons of asphalt pavement is reclaimed each year in the US, and over 95 percent of that total is reused or recycled with a savings of $1.8 billion per year for the taxpayer (9). Reclaimed Asphalt Pavement (RAP), also known as milled asphalt, is the generic term used to describe reprocessed pavement materials containing asphalt and aggregates. With the increasing cost of asphalt, the scarcity of quality aggregates, and the pressurizing need to preserve the environment, RAP has been favored over virgin materials in asphalt pavements (10). Fig. 5.4 represents the various possibilities for reusing RAP in road maintenance and reconstruction works (11).

The first sustained efforts to recover and reuse old asphalt paving materials in new asphalt pavements began in the 1970s in the US. Now, the use of processed RAP (see Fig. 5.5) as an aggregate substitute and as a portion of the binder to produce conventional Recycled Hot Mix (RHM) is a standard asphalt paving practice in nearly all 50 States in the US. Depending on State specifications, 10 to 30 percent RAP is typically permitted in HMA surface courses with some States permitting even higher percentages from approved RAP sources (12). Results of numerous field studies and standardized tests indicate that typical RAP can be used as construction fill without undue negative environmental consequences.

Since RAP aggregates are already coated with (aged) asphalt, it reduces the amount of binder required in the new construction. However, the level of interaction between aged and virgin asphalt binders has not been fully understood which has kept the percent of RAP in the recycled mixtures to relatively low values (10).

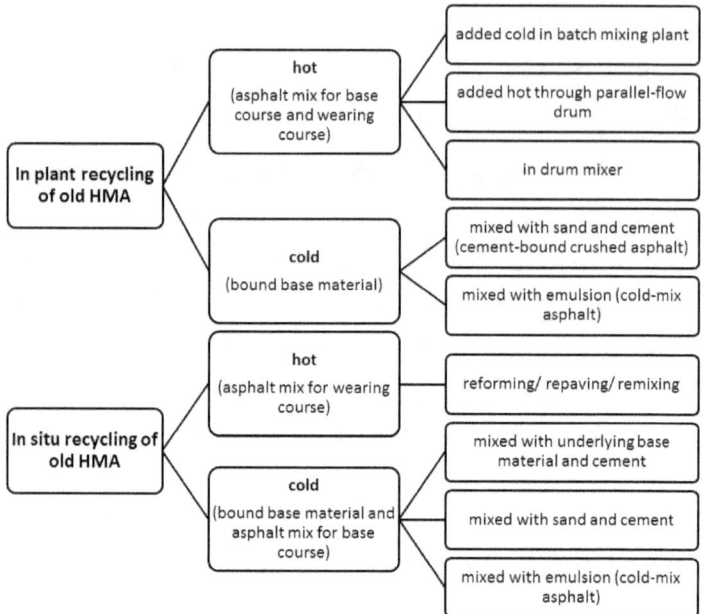

Fig. 5.4. Reclaimed Asphalt Pavement (RAP) recycling processes and end products [*Source:* RILEM (11)]

Fig. 5.5. Processing of RAP into fine and coarse aggregate stockpiles at an HMA production plant [*Source:* FHWA (13)]

Recycled Concrete Aggregate (RCA)

The production of construction aggregates is expected to increase to more than 2.5 billion tons per year by 2030. On the other hand, the construction waste produced from building demolition alone is estimated to be around 123 million tons per year (based on 2004 estimates). The growing environmental concerns related to depletion of natural resources and increasing landfills have led highway agencies and the aggregate industry to begin recycling concrete debris as an alternative aggregate. RCA is obtained from removing old PCC pavement, bridge structures/decks, sidewalks, curbs and gutters and crushed to a desired gradation after removing the steel (see Fig. 5.6). Commercial construction debris can also be a source of RCA, if unwanted materials like bricks, wood, steel, ceramics, and glass are removed (14).

A 2004 FHWA survey of State Transportation Agencies (STA) on the use of RCA revealed that concrete is routinely being recycled into the highway environment in the US and its principal application has been as a base material (14). Studies have shown that RCA used in base and subbase material of concrete pavements can provide performance comparable to or better than using virgin aggregate (14).

Fig. 5.6. Coarse (left) and Fine (right) RCA [*Source:* FHWA (15)]

The environmental and economic benefits associated with RCA are summarized in Fig. 5.7. Another sustainability benefit of RCA is that it has the potential to absorb a significant amount of CO_2 from the surrounding environment as the process of crushing concrete to produce RCA exposes areas of concrete that have not carbonated to atmospheric CO_2 (16).

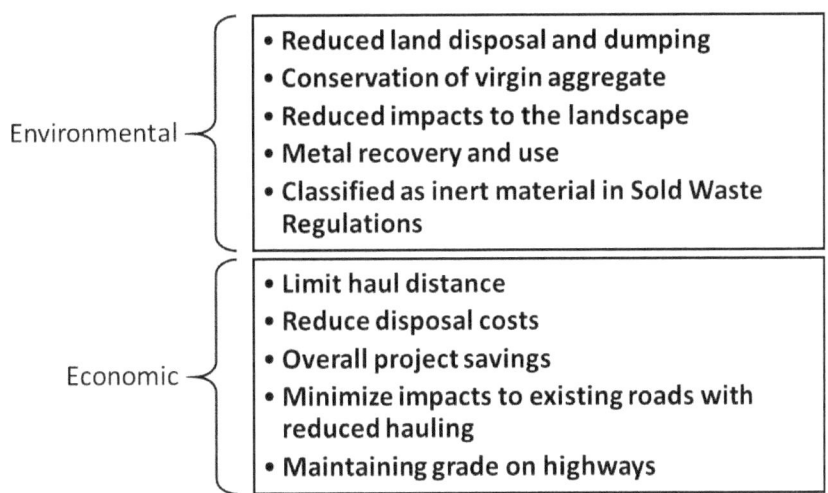

Fig. 5.7. Environmental and economic benefits of RCA [*Source:* FHWA (15)]

Cold In-Place Recycling (CIR)

Cold In-Place Recycling (CIR) involves cold milling the existing HMA surface; mixing the cold milled materials with emulsified asphalt or other modifiers to improve the properties of original HMA mix; screeding, spreading and compacting the recycled mixture in one continuous operation. Fig. 5.8 shows a typical CIR operation with a milling machine, a crusher and screening plant, and a mix-paver (17).

Two types of CIR are possible: (1) Full-Depth Reclamation (FDR) - the existing HMA pavement is milled at depths of 4 to 12 in., mixed with additional aggregate (as needed) to form a new layer; (2) Partial-Depth CIR – the existing HMA pavement surface is milled at depths 2 to 4 in.

(which includes only the bituminous portion of the pavement) to create base courses or even surface courses on highways with low to medium traffic volume (17,18). Generally, CIR refers to partial-depth CIR where the existing HMA surface alone is recycled.

The most significant advantages of using CIR are that treatment of pavements exhibiting a wide range of distresses; significant reduction in reflective cracking; structural improvement of pavements without changing horizontal and vertical alignments and without reconstructing shoulders; reduction in cost and energy, aggregate, and binder requirements compared to new construction; improvement of ride quality; and minimum hauling and air quality problems (17,18).

Fig. 5.8. Typical Cold-In-Place Recycling (CIR) Operation [*Source:* FHWA (17)]

Full-Depth Reclamation (FDR)

Full-Depth Reclamation (FDR) involves pulverizing and blending the distressed asphalt pavement section and a predetermined amount of underlying granular base material simultaneously to produce a stabilized base course (19,20). Different types of additives such as asphalt emulsions and chemical agents such as calcium chloride, Portland cement, fly ash, and lime, can be added to obtain an improved base. Pulverization, introduction of additive, shaping of the mixed material, compaction, and application of a surface or a wearing course are the five main steps involved in the FDR process.

FDR can treat pavements with deep rutting, load-associated cracks, non-load associated thermal cracks, and reflection cracks it is particularly recommended for pavements with base or subgrade problem. Four different types of in-place sizing and mixing operations are used at present for conducting FDR: multiple-step sequence, two-step sequence, single machine, and the single-pass equipment train (20).

The most significant advantages of using FDR include elimination of major distresses including alligator, transverse, longitudinal, and reflection cracking; improvement in ride quality; lower production and engineering cost as only a thin overlay or chip seal surfacing is required on most projects; conservation of materials and energy resources; and minimum hauling and air quality problems (20). Fig. captures the materials and energy savings achieved with FDR compared to a new base based on 1 mile of 24-foot wide 2-lane road with 6-inch base (21).

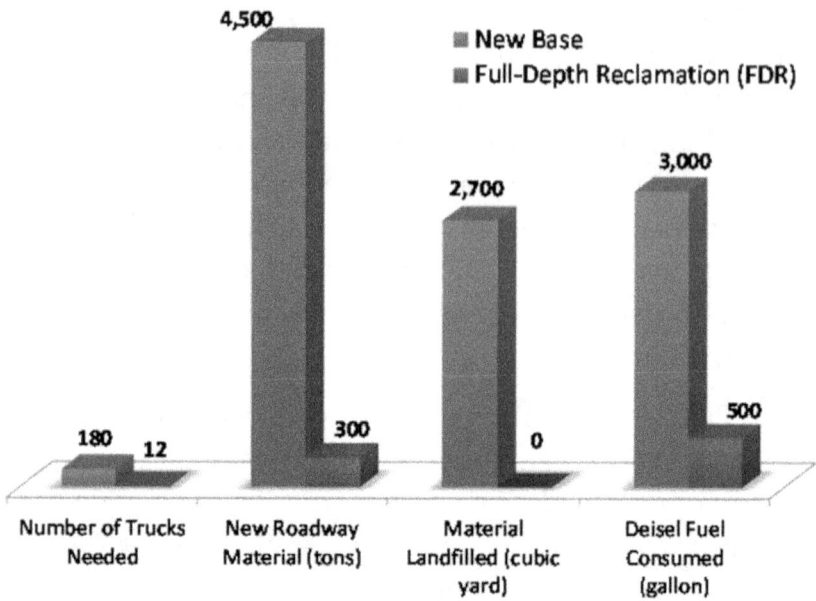

Fig. 5.9. Comparison of energy use and materials between Full-Depth Reclamation (FDR) and new base based on 1 mile of 24-foot wide 2-lane road with 6-inch base [*Source:* PCA (21)]

Rubblization

Rubblization is an in-place rehabilitation technique that involves breaking the existing deteriorating concrete pavement into pieces to form a high-quality crushed stone base layer for a new asphalt surface. The sizes of the broken pieces usually range from sand size to 3 in. at the surface (see Fig. 5.10) and 12 to 15 in. on the bottom part of the rubblized layer (22).

Rubblization is carried out either using a Resonant Pavement Breaker (RPB) or Multi-Head Breaker (MHB) and is applicable to all types of existing PCC pavements (JPCP, JRCP, and CRCP). Among the fractured slab techniques, it has gained increased acceptance and has become the most utilized procedure for addressing reflection cracking. The reported benefits of rubblization include economic savings to both users and agencies, speed of construction and construction during off-peak hours, elimination of ASR problems with the existing PCC, conservation of landfill space, reduced requirements of virgin aggregates and conservation of landfill space, and improved ride quality (22).

Fig. 5.10. Rubblized PCC surface [*Source:* M. Heitzman]

Beneficial Reuse of Industrial By-Products

It is estimated that some 600 million metric tons of industrial by-product source materials are produced annually which represents a sustainable solution to reduce the quantity of virgin material requirements in roadway projects as well as address energy and climate change issues (1). Select materials are often added to industrial by-products to generate materials suitable for highway construction applications with well-controlled and superior properties (23). The following are the major groups of industrial by-products used in roadway projects (see Fig. 5.11):

- Coal Combustion Products (CCPs)
  - o Fly ash
  - o Bottom ash
  - o Flue Gas Desulphurization (FGD) material
  - o Boiler slag
  - o Fluidized Bed Combustion (FBC) ash
  - o Cenospheres
- Iron and Steel Slag Products
  - o Blast Furnace Slag (BFS)
  - o Air-Cooled BFS
  - o Pelletized BFS
  - o Expanded BFS
  - o Granulated BFS or Ground Granulated BFS (GGBFS)
- Foundry Sands and Slags
- Tire Derived Rubber Materials
- Paper Production Materials

Apart from these materials, there are other waste and by-product materials which have been investigated for their potential use in pavement construction (24): baghouse fines, kiln dusts, mineral processing wastes, Municipal Solid Waste (MSW) incinerator ash, nonferrous slags, quarry by-products, roofing shingle scrap, sewage sludge ash, sulfate wastes, and waste glass (see Fig. 5.12).

APPLICATIONS	Coal Combustion Products	Iron & Steel Slag Products	Foundry Sands and Slags	Tire Derived Rubber Materials	Recycled Concrete Aggregates	Paper Production Materials
Asphalt Concrete Pavement						
Fine Aggregate or Filler	✓	✓	✓	✓	✓	
Coarse Aggregate	✓	✓	✓		✓	
Binder	✓	✓		✓		
PCC Pavement						
Replacement for PC	✓	✓				
Fine Aggregate	✓	✓	✓		✓	
Coarse Aggregate		✓	✓		✓	
Granular Bases						
Fine Aggregate	✓	✓	✓		✓	
Coarse Aggregate	✓	✓		✓	✓	
Subbase	✓	✓	✓	✓	✓	
Stabilized Bases						
Aggregate	✓	✓	✓		✓	
Binder	✓	✓				
Flowable Fill						
Binder	✓	✓				
Fine Aggregate	✓	✓	✓			
Coarse Aggregate	✓	✓		✓		
Other						
Concrete Barriers	✓	✓	✓		✓	
Sound Barriers	✓	✓	✓		✓	
Vegetation Control				✓		✓
Embankments						
Structural Fills						
Soil Stabilization						
Drainage Layers						
Compost						
Soil Conditioning						
Manufactured Soils						
Fertilization						

(Row groups at left: *Highways, Roads and Structures*, *Construction*, *Land Use*)

Fig. 5.11. Industrial By-Products and Applications [*Source: Industrial Resources Council (25)*]

CCBs are by-products generated from power plants when coal is burned to generate electricity. These include coal fly ash, bottom ash, boiler slag, and flue gas desulfurization gypsum. Fly ash (see Fig. 5.13) is most commonly used as a pozzalan in PCC applications (26). GGBFS, also known as slag cement, has been used in concrete for over a century in the US (27). Fly ash and GGBFS are the most commonly used Supplementary Cementitious Materials (SCMs) in concrete, either as an ingredient added at batching, or as a component of a blended cement or both (28).

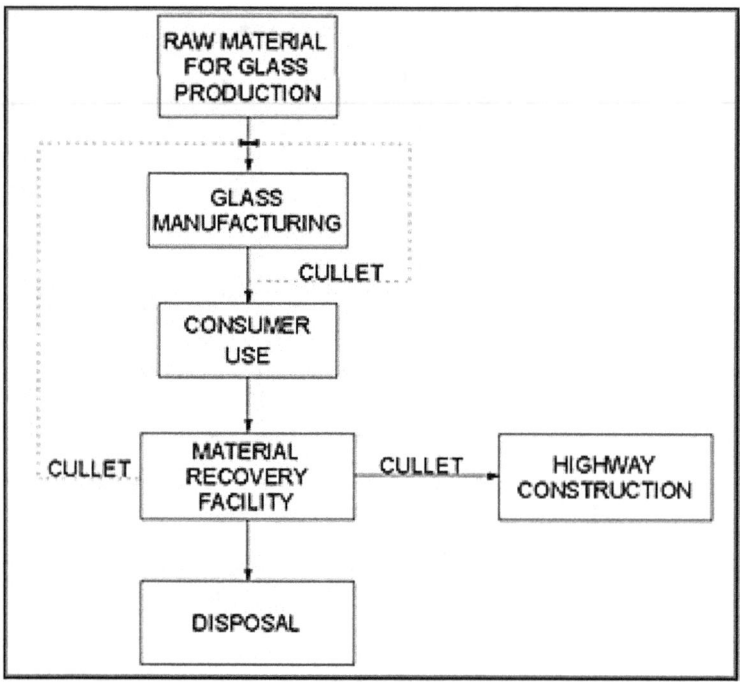

Fig. 5.12. Glass manufacturing, recovery, and reuse in the US [*Source:* FHWA (24)]

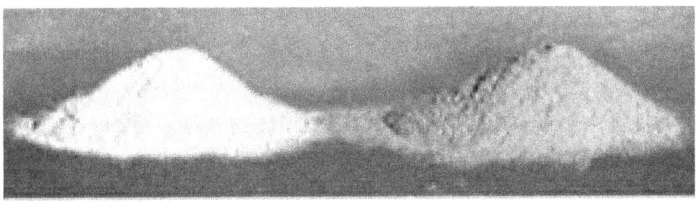

Fig. 5.13. Typical fly ash colors [*Source:* FHWA (26)]

Nearly 6 to 10 million tons of foundry sand are produced annually as by-products of the metal casting process from more than 3,000 foundries in the US. It can be used in many of the same ways as natural or manufactured sands in highway applications. In practice, it is primarily used in geotechnical bulk applications, such as embankments, site development fills and road bases (29).

Based on year 2003 estimates, more than 10 million tons of iron and steel slag sold in the US annually are used for road construction (30). Especially in areas where natural fine aggregate sources are scarce, its role as a substitute for natural aggregates has environmental, engineering and economic benefits (1).

In the US, roughly 300 million scrap tires (see Fig. 5.14) are generated annually of which 40 million tires still go to landfills or other land disposal options. Recycled tires can be cut into small pieces to produce Tire Derived Aggregate (TDA) for use in road construction. They can also be blended with asphalt as ground rubber to beneficially modify the asphalt properties for use in highway construction (1).

Fig. 5.14. Scrap tires [*Source:* FHWA (31)]

Useful Online Resources

Recycled Materials Resource Center (RMRC)

http://www.recycledmaterials.org/

http://www.rmrc.unh.edu/

US EPA's Website on Roadway Construction Using Industrial Materials

http://www.epa.gov/osw/conserve/rrr/imr/road.htm

US EPA's Industrial Materials Recycling Website

http://www.epa.gov/osw/conserve/rrr/imr/index.htm

FHWA's Office of Pavement Technology Recycling Homepage

http://www.fhwa.dot.gov/pavement/recycling/index.cfm

Industrial Resources Council

http://www.industrialresourcescouncil.org/

Green Highways Partnership – Recycle and Beneficial Reuse Group

http://www.greenhighwayspartnership.org/index.php?option=com_content&view=article&id=23&Itemid=22

American Coal Ash Association (ACAA)

http://www.acaa-usa.org/

Rubber Pavements Association

http://www.rubberpavements.org/

Peterson Asphalt Research Conference

http://www.petersenasphaltconference.org/

References

1. Green Highways Partnership (GHP). Green Highways Parternship (GHP) - Recycle and Beneficial Reuse Group [Internet]. 2010 [cited 2011 Feb 16];Available from: http://www.greenhighwayspartnership.org/index.php?option=com_content&view=article&id=23&Itemid=22

2. Japanese Ministry of the Environment. What are the 3Rs? [Internet]. Regional 3R Forum in Asia. [cited 2011 Feb 16];Available from: http://www.env.go.jp/recycle/3r/en/forum_asia/regional _3r/index.html

3. Reid JM, Chandler JWE, Schiavi I, Hewitt AP. Sustainable Highways: A Short Guide [Internet]. London, UK: Department for Transport (UK Roads Board); 2008 [cited 2011 Mar 14]. Available from: http://www.handapp.co.uk/web/Sustainable%20highwa ys%20A%20short%20guide.pdf

4. FHWA. FHWA Policy Memos: FHWA Recycled Materials Policy [Internet]. 2002 Feb 7 [cited 2011 Feb 16];Available from: http://www.fhwa.dot.gov/legsregs/directives/policy/rec matpolicy.htm

5. Van Dam T, Taylor P. Building Sustainable Pavements with Concrete: Briefing Document [Internet]. 2009 Aug [cited 2011 Feb 6];Available from: http://www.cproadmap.org/publications/sustainability_ briefing.pdf

6. SCI's Construction Materials Group. Conference Announcement: Recycling and Re-using Asphalt [Internet]. 2011 [cited 2011 Feb 16];Available from: http://www.highwaysmaintenance.com/Rcycling/SOCI %20Mar_2011_Recycling_Reusing_Asphalt.pdf

7. FHWA. Recycled Materials in European Highway Environments: Uses, Technologies, and Policies [Internet]. Washington D.C.: Federal Highway Administration (FHWA); 2000 [cited 2011 Feb 18].

Available from:
http://international.fhwa.dot.gov/pdfs/recycolor.pdf

8.   US EPA. Using Recycled IndustrialMaterials in Roadways [Internet]. Washington D.C.: US Environmental Protection Agency (EPA); 2009 [cited 2011 Feb 16]. Available from: http://www.epa.gov/osw/conserve/rrr/imr/pdfs/roadways.pdf

9.   APA. The Role of Asphalt in Livable Communities [Internet]. Lanham, MD: Asphalt Pavement Alliance (APA); 2010 [cited 2011 Feb 22]. Available from: http://asphaltroads.org/images/documents/livablecommunities_whitepaper_september2010.pdf

10. Al-Qadi IL, Elseifi MA, Carpenter SH. Reclaimed Asphalt Pavement – A Literature Review [Internet]. University of Illinois at Urbana-Champaign, Urbana, IL: Illinois Center for Transportation (ICT); 2007 [cited 2011 Feb 16]. Available from: http://ict.illinois.edu/publications/report%20files/FHWA-ICT-07-001.pdf

11. De Bock L, Gonçalves A. Recycled Asphalt Pavement [Internet]. In: Hendriks CF, Janssen GMT, Vázquez E, editors. Use of Recycled Materials - Final Report of RILEM TC 198-URM. Bagneux, France: RILEM: International Union of Laboratories and Experts in Construction Materials, Systems and Structures; 2005 [cited 2011 Feb 24]. p. 45-51.Available from: http://www.rilem.net/repContents.php?rep=rep030

12. RMRC. User Guidelines for Waste and Byproduct Materials in Pavement Construction [Internet]. 2008 Jul [cited 2011 Feb 16];Available from:

http://www.recycledmaterials.org/tools/uguidelines/index.asp

13. FHWA. Asphalt Pavement Recycling with Reclaimed Asphalt Pavement (RAP) [Internet]. 2010 Jul [cited 2011 Feb 16];Available from: http://www.fhwa.dot.gov/pavement/recycling/rap/index.cfm

14. FHWA. Recycled Concrete Study Identifies Current Uses, Best Practices. Focus [Internet]. 2004 Apr [cited 2011 Feb 16];FHWA-HRT-04-024. Available from: http://www.fhwa.dot.gov/publications/focus/04apr/01.cfm

15. FHWA. Transportation Applications Of Recycled Concrete Aggregate [Internet]. Washington D.C.: Federal Highway Administration (FHWA); 2004 [cited 2011 Feb 16]. Available from: http://www.rmrc.unh.edu/Research/tools/RCAREPORT.pdf

16. PCA. Recycled Aggregates [Internet]. Recycled Aggregates for Use in Concrete. [cited 2011 Feb 16];Available from: http://www.cement.org/tech/cct_aggregates_recycled.asp

17. FHWA. Chapter 1. Introduction to Pavement Recycling [Internet]. Washington, D.C.: Federal Highway Administration (FHWA); 2007 [cited 2011 Feb 24]. Available from: http://www.fhwa.dot.gov/pavement/recycling/98042/01.cfm

18. Salomon A, Newcomb DE. Cold In-Place Recycling: Literature Review and Preliminary Mixture Design Procedure. Minneapolis, MN: Minnesota Department of Transportation; 2000.

19. ARRA. Full Depth Reclamation: A Century of Advancement for the New Millennium [Internet]. Annapolis, Maryland: Asphalt Recycling & Reclaiming Association (ARRA); 2001 [cited 2011 Feb 24]. Available from: http://www.pavementpreservation.org/toolbox/links/arrafull.pdf

20. FHWA. Chapter 16. Full Depth Reclamation (Construction Methods And Equipment) [Internet]. Pavement Recycling Guidelines for State and Local Governments: Participant's Reference Book. 2007 Mar [cited 2011 Feb 24];Available from: http://www.fhwa.dot.gov/pavement/recycling/98042/16.cfm

21. PCA. Full-Depth Reclamation (FDR) - Recycling failed pavements [Internet]. Portland Cement Association (PCA). 2011 Feb 24 [cited 2011 Feb 24];Available from: http://www.cement.org/pavements/pv_sc_fdr.asp

22. TRB. Rubblization ofPortland CementConcrete Pavements [Internet]. Washington, D.C.: Transportaion Research Board (TRB); 2006 [cited 2011 Feb 24]. Available from: http://www.gahotmix.com/tech-rubblization20060101.pdf

23. Edil TB. Green Highways: Strategy for Recycling Materials forSustainable Construction Practices [Internet]. In: Proceedings of the Seventh International Congress on Advances in Civil Engineering. Istanbul,

Turkey: Yıldız Technical University; 2006 [cited 2011 Apr 2]. p. 20.Available from: http://www.resourcesaver.org/ewebeditpro/items/O130 F10111.pdf

24. FHWA. User Guidelines for Waste and Byproduct Materials in Pavement Construction - Revised [Internet]. Washington, D.C.: Federal Highway Administration (FHWA); 2011 [cited 2011 Mar 15]. Available from: http://www.fhwa.dot.gov/publications/research/infrastr ucture/pavements/97148/004.cfm#content

25. IRC. Industrial Materials Applications Matrix [Internet]. Industrial Resources Council (IRC). 2008 [cited 2011 Feb 25];Available from: http://www.industrialresourcescouncil.org/Application s/IndustrialMaterialsMatrix/EMatrix/tabid/409/Default. aspx

26. FHWA. Fly Ash Facts for Highway Engineers [Internet]. 2003 [cited 2011 Feb 22]. Available from: http://www.fhwa.dot.gov/Pavement/recycling/fafacts.p df

27. SCA. What is Slag Cement? [Internet]. Slag Cement Association (SCA). 2011 [cited 2011 Feb 23];Available from: http://www.slagcement.org/

28. PCA. What are supplementary cementitious materials (SCMs) and how do they differ from blended cements? [Internet]. Concrete Technology FAQs. 2011 [cited 2011 Feb 23];Available from: http://www.cement.org/tech/faq_scms.asp

29. Foundry Industry Recycling Starts Today (FIRST). Foundry Sand Factsfor Civil Engineers [Internet]. Washington D.C.: Federal Highway Administration (FHWA); 2004 [cited 2011 Feb 22]. Available from: http://isddc.dot.gov/OLPFiles/FHWA/011435.pdf

30. NSA. Iron and Steel Making Slag –Environmentally ResponsibleConstruction Aggregates [Internet]. Pleasant Grove, Utah: National Slag Association (NSA); 2003 [cited 2011 Feb 18]. Available from: http://www.nationalslag.org/archive/nsa_202-3_environmental_tech_bulletin.pdf

31. Harrington J. Recycled Roadways. Public Roads [Internet]. 2005 [cited 2011 Apr 2];68(4). Available from: http://www.fhwa.dot.gov/publications/publicroads/05jan/02.cfm

Chapter 6:
Long-Life Pavements

An important aspect of sustainable development of roadways is to construct long-lasting pavements that will minimize the cost and frequency of maintenance and rehabilitation activities as well as user costs associated with traffic delays, operating costs vehicle repairs over the pavement life cycle.

Long-life pavements contribute to improved highway safety, congestion mitigation, and resource conservation leading to reduced environmental impacts by requiring less frequent repair, rehabilitation, and reconstruction. The Mechanistic-Empirical Pavement Design Guide (MEPDG) for New and Rehabilitated Pavement Structures is expected to help achieve reliable, cost-effective, and sustainable long-life pavement designs as it incorporates improved pavement design technology that allows optimization of many key pavement design features (1).

Perpetual Pavements: Long-Life Asphalt Pavements

The development of the "Perpetual Pavement" design concept was introduced in 2000 by the Asphalt Pavement Alliance (APA) (2). According to the APA, a Perpetual Pavement is "an asphalt pavement designed and built to last longer than 50 years without requiring major structural rehabilitation or reconstruction, and needing only periodic surface renewal in response to distresses confined to the top of the pavement." (3). The idea of Perpetual Pavements has evolved from the successes seen in the long-lasting performance of full-strength and deep-strength asphalt pavements in resisting bottom-up fatigue cracking and deep structural rutting, the two most devastating asphalt pavement distresses (3,4). In fact, Perpetual Pavements have been in existence since the 1960s under different names such as thick-asphalt pavements, long-lasting asphalt pavements, extended life HMA pavements, etc (3).

The premise behind Perpetual Pavement design approach is that if pavement critical responses such as deflections, stresses, and strains are kept below certain thresholds (*limiting pavement responses* or *endurance limit*), then deep seated structural distresses could be avoided almost indefinitely (3-6). When surface-initiated distresses do occur and propagate downward, they could be corrected by periodic maintenance and rehabilitation activities before they affect the bulk of the pavement structure. This results in lower life-cycle costs as the bulk of the pavement structure remains almost intact throughout the design life (7). Although the initial construction costs for Perpetual Pavements are at least over 20% than that of conventional pavements, their long-term benefits are expected to outweigh this (8).

In addition to providing a long-life structure, the idea of Perpetual Pavements came into existence also to prevent over-design resulting from the use of empirically-based *1993 AASHTO Guide for the Design of Pavement Structures* (9) which uses an approach of "ever-increasing pavement thickness with traffic volume". Over-designed pavements require large quantities of construction material per lane-mile beyond what is needed leading to unsustainable use of resources in addition to being extravagant from a cost standpoint (3,10).

Perpetual Pavements (see Fig. 6.1) are designed for the heaviest anticipated traffic loads by mechanistically identifying the critical pavement responses which are low enough (*limiting pavement responses*) to avoid the initiation of fatigue cracking or rutting in the pavement structure (5,10-12). Although most Perpetual Pavements are designed to last for 50 years, periodic resurfacing is generally needed within 20 years to improve the surface characteristics, reduce noise, and mitigate surface cracking. Apart from the design itself, quality materials and construction practices and timely preservation activities are critical to long-life pavements (3,4).

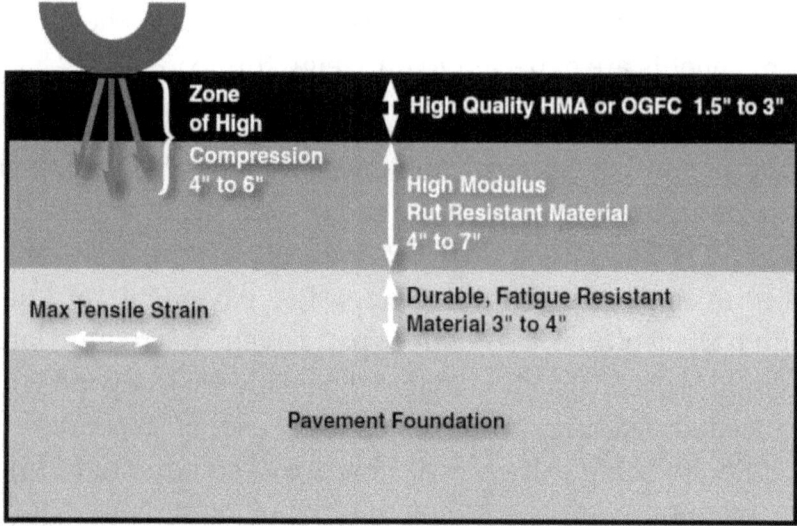

Fig. 6.1. Perpetual Pavement Design Concept [*Source:* APA (2,3)]

Long-life pavements are generally associated with high initial construction costs, but lower life-cycle costs resulting from reduced maintenance and rehabilitation activities (13). To determine whether long-life pavements are economically justifiable on high-traffic roads, the OECD initiated the Long Life Pavements project in 2002 (14). During the first phase, an economical evaluation of long life surface (or wearing) course was performed and it was concluded that "long life pavement surfacing costing around three times that of traditional wearing courses could be economically viable, assuming an expected life of 30 years and an Annual Average Daily Traffic (AADT) of 80,000 or more" (14). The third phase of this study began in 2009 to conduct field trials of road sections constructed with epoxy asphalt and high-performance cementitious materials and is currently in progress (15).

Perpetual Pavement design has been successfully used in a variety of applications including high-volume pavements (5,8,11,16), low-volume pavements (7), high-modulus pavements (17), and the rehabilitation of flexible and rigid pavements (18). The latest version of the MEPDG software accommodates multiple pavement layers and has an analysis period of more than 50 years providing an ideal mechanistic-empirical tool for modeling Perpetual Pavements (19).

The following benefits have been attributed to the use of Perpetual Pavements that are in line with the principles of sustainable development (8,16):

- ✓ They provide efficient design, thus eliminating costly over-designed pavements
- ✓ They offer high structural capacity for heavy truck loads and high traffic volume
- ✓ They eliminate reconstruction costs by limiting pavement responses within pavement's structural capacity
- ✓ They reduce user costs associated with rehabilitation or maintenance delays which can be significant on heavily-trafficked highways
- ✓ They promote conservation of non-renewable resources like aggregates and asphalt
- ✓ They diminish the energy costs of in-service pavements
- ✓ They reduce pavement life-cycle costs with minimal or no major structural rehabilitation activities

Long-Life Concrete Pavements

In the past, majority of the concrete pavements in the US interstate and primary systems were designed and constructed to provide low-maintenance service lives of 20 to 25 years. However, many PCC pavements often last in excess of 35 years. The NCHRP Project 1-32 survey (20) conducted in 1997 revealed that 7 States in the US had PCC pavement design lives of at least 35 years and 2 of them had lives of at least 40 years (21). Also, a number of very old PCC pavements still remain in service in the US (see Fig. 6.2) indicating that long-life concrete pavements have been quite attainable for a long time (22).

Since 1997, long-life concrete pavement design has evolved with recent advances in design, construction, concrete materials technology, and testing procedures. For instance, the Minnesota DOT has set a 60-year PCC pavement design life standard (23). The California Department of Transportation (Caltrans) requires use of a 40-year service life for designing pavements along a high-volume, urban corridor where traffic disruptions and user delays can be acute with frequent or extended lane closures for rehabilitation (22). Concrete pavements designed to last for 40+ years are becoming an established practice in the US (22).

Recent definitions of long-life concrete pavements also include other desirable pavement characteristics such as reduced potential for cracking, faulting, and spalling and not exhibiting premature construction and materials-related distress. Also, long-life concrete pavements are required to maintain desirable ride and surface texture characteristics with minimal intervention (22).

Under the FHWA International Technology Scanning Program, a team of experts conducted a scan of European and Canadian design philosophies, material requirements, construction procedures, and maintenance strategies used to build long-life concrete pavements in these countries. Based on the scanning study, the team recommended the use of two-lift construction, using better-quality materials in pavement subbases, paying careful attention to cement and concrete mixture properties, and using geotextile interlayer to prevent concrete slabs from bonding to the cement-treated base for implementation in the US in designing long-life concrete pavements (24).

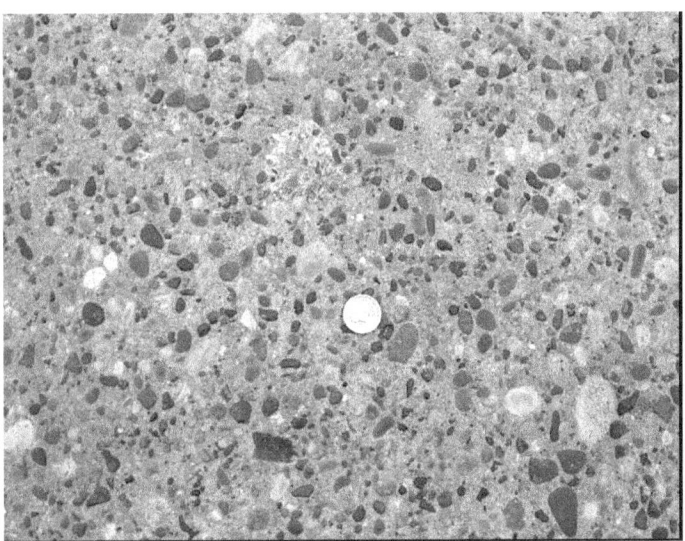

Fig. 6.2. Exposed aggregate surface of a 50-year old PCC pavement that is still performing well in City of Bettendorf, Iowa [*Source:* (25)]

Useful Online Resources

Long-Life Concrete Pavements

FHWA's Publications Website on Long-Life Concrete Pavements

http://www.fhwa.dot.gov/pavement/longlife/index.cfm

The Concrete Pavement Road Map: Long-Term Plan for Concrete Pavement Research and Technology

http://www.cptechcenter.org/publications/task15/task15.cfm

http://www.cproadmap.org/research/track.cfm?trackID=8

International Conference on Long-Life Concrete Pavements (Oct. 25-27, 2006)

http://www.fhwa.dot.gov/pavement/concrete/2006conf.cfm

Long-Life Concrete Pavements in Europe and Canada: FHWA International Scan Tour (Aug. 2007)

http://international.fhwa.dot.gov/pubs/pl07027/

Perpetual Pavements

PerRoad: APA's Mechanistic-Empirical Perpetual Pavement Design Software http://asphaltroads.org/why-asphalt/perpetual-pavement.html

International Conference on Perpetual Pavement (Sep. 13-15, 2006)

http://www.ohio.edu/icpp/Cnference-Proceedings.cfm

References

1. NCHRP. Guide for Mechanistic-Empirical Design of New and Rehabilitated Pavement Structures [Internet]. Washington D.C.: National Cooperative Highway Research Program (NCHRP); 2004 [cited 2011 Feb 7]. Available from: http://pubsindex.trb.org/view.aspx?id=703699

2. Newcomb DE, Buncher M, Huddleston IJ. Concepts of Perpetual Pavements. In: Perpetual Bituminous Pavements. Transportation Research Board; 2001. p. 4-11.

3. Newcomb DE, Willis R, Timm DH. Perpetual Asphalt Pavements: A Synthesis. Lanham, MD: Asphalt Pavement Alliance (APA); 2010.

4. Prowell B, Brown ER, Anderson RM, Daniel JS, Krishna Swamy A, Von Quintus HV, et al. Validating the fatigue endurance limit for hot mix asphalt [Internet]. Washington D.C.: Transportation Research Board; 2010 [cited 2011 Feb 2]. Available from: http://onlinepubs.trb.org/onlinepubs/nchrp/nchrp_rpt_646.pdf

5. Nunn ME, Brown A, Weston D, Nicholls JC. Design of Long-Life Flexible Pavements for Heavy Traffic. Berkshire, United Kingdom: Transport Research Lab (TRL); 1997.

6. Ferne BW, Nunn M. The European Approach to Long Lasting AsphaltPavements – A State-of-the-Art Review by ELLPAG. In: International Symposium on Design and Construction of Long Lasting Asphalt Pavements. Auburn, Alabama: International Society for Asphalt Pavements (ISAP); 2004. p. 87-101.

7. Muench ST, Mahoney JP, Wataru W, Chong L, Romanowski J. Best Practices for Long-Lasting Low-Volume Pavements. Journal of Infrastructure Systems. 2007;13(4):311-320.

8. Timm DH, Newcomb DE. Perpetual pavement design for flexible pavements in the US. International Journal of Pavement Engineering. 2006;7(2):111–119.

9. American Association of State Highway and Transportation Officials. AASHTO guide for design of pavement structures. Washington D.C.: AASHTO; 1993.

10. Huber G, Andrewski D, Gallivan V. Design and Construction of Highways for Very Heavy Trucks [Internet]. In: 2009 International Conference on Perpetual Pavement. Columbus, Ohio: 2009. Available from: http://www.ohio.edu/icpp/upload/Program.pdf

11. Thompson MR, Carpenter SH. Design principles for long lasting HMA pavements. In: Proceedings of International Symposium on Design and Construction

of Long Lasting Asphalt Pavements. Auburn University, Alabama: 2004. p. 365–384.

12. Timm DH. PerRoad. Auburn University, Alabama: 2008.

13. Muench ST, Anderson JL, Hatfield JP, Koester JR, Soderlund M. Greenroads Rating System v1.5. Seattle, WA: University of Washington;

14. OECD Staff. Economic evaluation of long-life pavements: PHASE 1 [Internet]. OECD Publishing; 2005 [cited 2011 Feb 1]. Available from: www.internationaltransportforum.org/Pub/pdf/05Pave mentI.pdf

15. FEHRL. Long Life Pavements (phase3)- Field Trials - call for participants [Internet]. FEHRL: news. 2009 Jan [cited 2011 Feb 2];Available from: http://www.fehrl.org/?m=33&a=content&id=149

16. Walubita LF, Liu W, Scullion T, Leidy J. Modeling Perpetual Pavements Using the Flexible Pavement System (FPS) Software [Internet]. In: Transportation Research Board Annual Meeting 2008. 2008 [cited 2011 Feb 2]. p. 14.Available from: http://pubsindex.trb.org/view.aspx?id=848479

17. Lee HJ, Lee JH, Park HM. Performance evaluation of high modulus asphalt mixtures for long life asphalt pavements. Construction and Building Materials [Internet]. 2007 May [cited 2011 Feb 2];21(5):1079-1087. Available from: http://www.sciencedirect.com/science/article/B6V2G-4JRVFVS-1/2/7189fd7a08565693ecebcb496b4ceae2

18. SHRP. Using Existing Pavement In Place and Achieving Long Life [Internet]. SHRP 2 Project - R23. [cited 2011 Feb 2];Available from: http://144.171.11.40/cmsfeed/TRBNetProjectDisplay.asp ?ProjectID=2174

19. Walubita LF. Computational Modeling of Perpetual Pavements Using the MEPDG Version 0.910 Software [Internet]. In: Proceedings of the 4th International SIIV Congress. Palermo, Italy: Societa Italiana Infrastrutture Viarie (SIIV); 2007 [cited 2011 Feb 2]. p. 15.Available from: http://sed.siiv.it/documenti/63_2848_20080107222544.pd f

20. NCHRP. NCHRP Project 1-32: Pavement Design Tools [Internet]. 1998 [cited 2011 Feb 6];Available from: http://pubsindex.trb.org/view.aspx?id=655412

21. Federal Highway Administration (FHWA). FHWA Sustainable Highways Self-Evaluation Tool [Internet]. 2011 [cited 2011 Feb 6];Available from: https://www.sustainablehighways.org/1/home.html

22. FHWA. Long-Life Concrete Pavements:Best Practices and DirectionsFrom the States [Internet]. Washington D.C.: Concrete Pavement Technology Program (CPTP), FHWA; 2007 [cited 2011 Feb 6]. Available from: http://www.fhwa.dot.gov/pavement/concrete/pubs/070 30/07030.pdf

23. Burnham T, Izevbekhai B, Rangaraju PR. The Evolution of High-Performance Concrete Pavement Design in Minnesota [Internet]. In: Tayabji S, Lim S, editors. Proceedings of the International Conference on Long-Life Concrete Pavements. Chicago, IL: FHWA; 2006. p.

25–27.Available from:
http://www.fhwa.dot.gov/pavement/concrete/2006conf.
cfm

24. Hall K, Dawood D, Vanikar S, Tally R, Cackler ET, Correa A, et al. Long-Life Concrete Pavements in Europeand Canada [Internet]. Washington D.C.: Federal Highway Administration (FHWA); 2007 [cited 2010 Feb 6]. Available from: http://international.fhwa.dot.gov/pubs/pl07027/pl07027. pdf

25. Ceylan H, Cable JK, Gopalakrishnan K. Defining the Attributes of Well-Performing. Long-Lasting Jointed Portland Cement. Concrete Pavements [Internet]. In: Tayabji S, editor. Proceedings of the International Conference on Long-Life Concrete Pavements. Rosemont, IL: Federal Highway Administration (FHWA); 2006 [cited 2011 Mar 31]. Available from: http://www.ctre.iastate.edu/research/detail.cfm?projectI D=-801793574.

Chapter 7:
Warm-Mix Asphalt (WMA)

What is WMA?

Warm-Mix Asphalt (WMA) is a generic term for technologies that allow reduction in temperatures at which the conventional asphalt mixes are produced and placed in the road (1). Traditional Hot Mix Asphalt (HMA) is heated at temperatures in excess of 300 ºF to allow the asphalt binder viscous enough to coat the aggregates during production, have workability while paving and durability when the pavement is exposed to traffic. When using HMA mixtures containing polymer-modified asphalt binders and crumb rubber asphalt binders, even higher temperatures are used (2).

However, with the use of WMA, temperature reductions of 50 to 100 ºF have been documented (3). Thus, an immediate benefit of producing WMA is the reduction in energy consumption required to heat traditional HMA by burning fuels. This naturally leads to other added benefits such as reduced GHG emissions generated from burning fuels and healthy construction work environment resulting from reduced fumes and odors (1). It has been reported that a reduction in temperature by 50 °F results in an emission reduction by 50%, as a rule of thumb (4).

In 2007, a team of experts from the FHWA International Technology Scanning Program conducted a scan tour of various WMA technologies found in Europe. The scan team reported that WMA technologies have been used with all types of asphalt mixtures, including dense-graded asphalt, Stone Matrix Asphalt (SMA), and porous asphalt. Based on laboratory and short-term field performance data, WMA mixes seemed to provide similar or better performance than HMA (5).

Other Low-Temperature Asphalt Mixes

Currently, there are more than fifteen different patented WMA technologies that are either foaming based or based on chemical or organic additive(s) (6). Actually, WMA comes under the broad category of low-temperature asphalt mixes which have been in existence in some form or other both in the US and Europe (6). But, the Kyoto Protocol and other current and impending environmental regulations regarding CO_2 emissions have generated renewed interest in these low-temperature mixes which mixes include Cold Mix Asphalt (CMA), Half-Warm Mix Asphalt (HWMA), and WMA (4,7). A comparison of these mixes with traditional HMA in terms of temperature regimes at which they are produced and fuel usage is shown in Fig. 7.1. A timeline with important milestones in the development of low-temperature asphalt technologies is shown in Fig. 7.2.

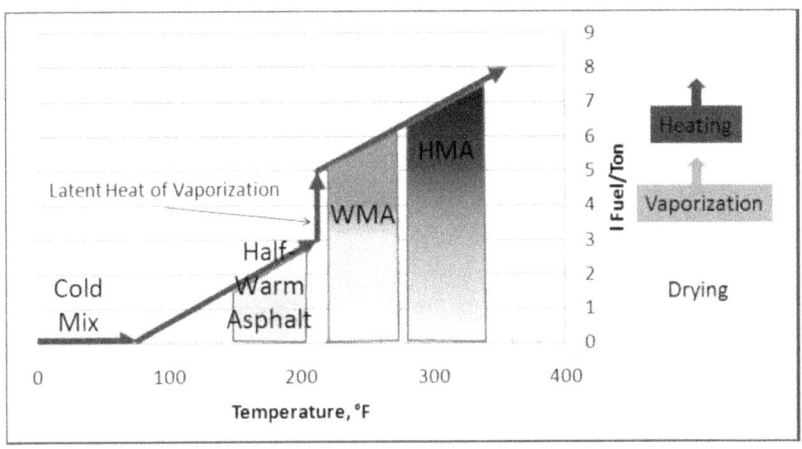

Fig. 7.1. Asphalt mixtures temperature regimes and fuel usage [*Source*: FHWA (8); *Note*: Classification by temperature range, temperatures, and fuel usage are approximations.]

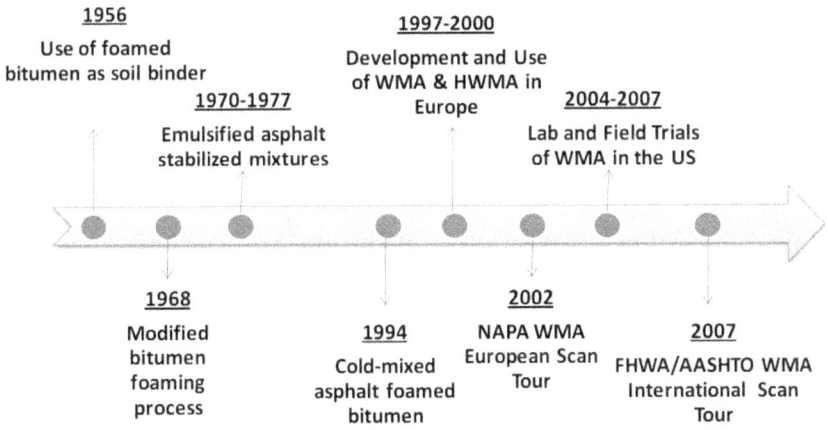

Fig. 7.2. A brief history of low-temperature asphalt technologies [*Source:* (6)]

WMA Research

A NCHRP project (9) was initiated in 2007 to develop a mix design method for WMA in the form of a draft AASHTO-recommended practice which could be used by engineers and technicians in the public and private sectors. This project has been completed and the final report (NCHRP Report 691) is scheduled to be released by mid 2011.

Another NCHRP project (2) was initiated in 2008 to examine the engineering properties, emissions, and long-term field performance of WMA technologies. This project has entered its second phase of its research and is expected to specifically address key issues related to WMA's rutting potential and its susceptibility to moisture damage which are deemed as some of the potential engineering challenges to widespread implementation of this technology (2). Apart from the two NCHRP projects, several ongoing research studies funded by State agencies and private industries are seriously investigating the different aspects of this new technology.

WMA Sustainability Benefits

A number of studies have reported that WMA has significant potential to increase construction efficiency, conserve natural resources, and advance environmental stewardship owing to its vastly reduced energy requirements and emissions. Compared to standard HMA, expected reductions of up to

30% in energy consumption and 30-40% in CO_2 emissions are reported (5,6). A summary of major WMA benefits due to its lower production and placement temperatures is captured in Fig. 7.3.

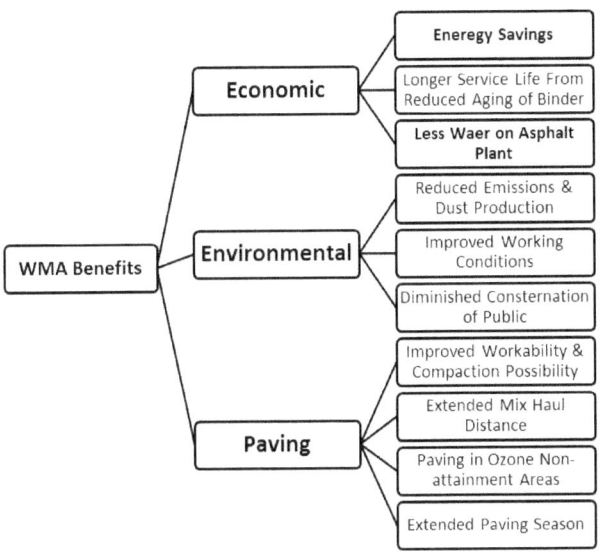

Fig. 7.3. Major benefits associated with WMA owing to its lower production and placement temperatures [*Source:* (5,6)]

Useful Online Resources

WMA Technical Working Group

http://www.warmmixasphalt.com/

FHWA's Website on WMA Technologies and Research

http://www.fhwa.dot.gov/pavement/asphalt/wma.cfm

WMA European Practice: FHWA's International Technology Scan Tour http://international.fhwa.dot.gov/pubs/pl08007/

References

1. FHWA. Warm Mix Asphalt Technologies and Research [Internet]. FHWA Office of Pavement Technology. [cited 2011 Feb 2];Available from: http://www.fhwa.dot.gov/pavement/asphalt/wma.cfm

2. NCHRP. NCHRP 09-47: Engineering Properties, Emissions, and Field Performance of Warm Mix Asphalt Technologies [Internet]. [cited 2011 Feb 2];Available from: http://144.171.11.40/cmsfeed/TRBNetProjectDisplay.asp?ProjectID=1625

3. WMA Technical Working Group (TWG). Warm Mix Asphalt (WMA) [Internet]. [cited 2011 Feb 2];Available from: http://www.warmmixasphalt.com/

4. EAPA. Low Temperature Asphalt [Internet]. 2005 Jun [cited 2011 Feb 2];Available from: http://www.eapa.org/usr_img/position_paper/low_temperature_asphalt2005.pdf

5. D'Angelo J, Harm E, Bartoszek J, Baumgardner G, Corrigan M, Cowsert J, et al. Warm-Mix Asphalt: European Practice [Internet]. Washington D.C.: Office of International Programs, Federal Highway Administration (FHWA); 2008 [cited 2011 Feb 2]. Available from: http://www.warmmixasphalt.com/submissions/68_20080223_FHWA-PL-08-007.pdf

6. Chowdhury A, Button JW. A Review of Warm Mix Asphalt [Internet]. College Station, Texas: Texas Transportation Institute, the Texas A&M University

System; 2008. Available from:
http://swutc.tamu.edu/publications/technicalreports/473
700-00080-1.pdf

7.   Miller TD, Bahia HU. Sustainable Asphalt Pavements:
      Technologies, Knowledge Gaps and Opportunities - A
      White Paper [Internet]. 2009 Feb [cited 2011 Feb
      2];Available from: http://uwmarc.wisc.edu/files/MARC-
      Sustainable-Asphalt-Pavements-white-paper.pdf

8.   Prowell B. Warm Mix Asphalt: The International
      Technology Scanning Program Summary Report
      [Internet]. Washington D.C.: Federal Highway
      Administration (FHWA); 2007 [cited 2011 Feb 3].
      Available from:
      http://international.fhwa.dot.gov/pubs/wma/wma.pdf

9.   NCHRP. NCHRP 09-43: Mix Design Practices for Warm
      Mix Asphalt [Internet]. [cited 2011 Feb 2];Available
      from:
      http://apps.trb.org/cmsfeed/TRBNetProjectDisplay.asp?
      ProjectID=977

Chapter 8:
Permeable Pavements

Urbanization and Stormwater Runoff

Rainwater purification occurs naturally as it percolates through soil on its way towards surface waters and groundwater aquifers. However, urbanization replaces more of the natural landscape with large areas of impervious pavement surfaces or roofs which directly carry the rainwater or melted snow through storm drainage systems into waterways. Thus, the stormwater which would have been naturally absorbed into the ground with beneficial environmental effects are prevented by these infrastructural elements which contribute to debris and chemicals in the stormwater runoff (1). Also, stromwater is increasingly becoming recognized as a valuable consumptive commodity that should no longer be discharged wantonly as a "waste product".

Fig. 8.1 depicts the effect of impervious surfaces on natural hydrologic cycle processes. The highway runoff has been found to contain certain priority pollutants, such as hydrocarbons, metals, nutrients, and pathogens, resulting from traffic using the facility (2). If the rapidly flowing stormwater is discharged untreated into storm drains, sewer systems, and ditches, it can cause downstream flooding, stream bank erosion, habitat destruction, combined sewer overflows, infrastructure damage, and the accumulated pollutants could adversely affect water quality (3). Recent research has consistently reported a strong relationship between the percentage of impervious cover in a watershed and the health of the receiving stream (3).

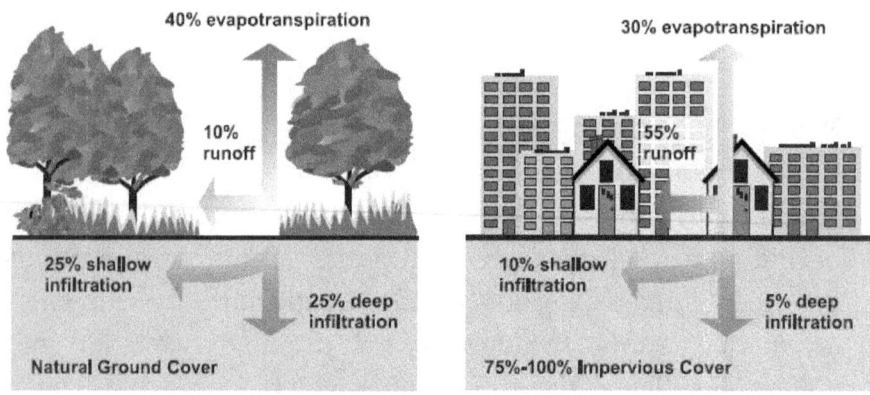

Fig. 8.1. Illustration of relationship between impervious surfaces and surface runoff [*Source:* US EPA]

Low-Impact Development (LID) and Wet-Weather Green Infrastructure

The US EPA proposes several approaches to decrease the stormwater impact from pavements and buildings on the environment, including Low Impact Development (LID), Wet Weather Green Infrastructure, and best practices. LID is defined as "a sustainable landscaping approach that can be used to replicate or restore natural watershed functions and/or address targeted watershed goals and objectives." And, the "infrastructure associated with stormwater management and low impact development that encompasses approaches and technologies to infiltrate, evapotranspire, capture, and reuse stormwater to maintain or restore natural hydrologies" is referred to as wet weather green infrastructure (3).

Stormwater BMPs

The industries and transportation agencies have their own Best Management Practices (BMPs) or best practices for stormwater management which include both structural and non-structural categories (4,5). While structural BMPs aim to mitigate stormwater-related impacts after they have occurred, non-structural best practices take broader planning and design approaches to prevent stormwater runoff from the site.

The non-structural BMPs include measures to preserve existing natural features and systems and manage stormwater as close to the source as possible. These include landscaping and vegetative practices, pesticide and fertilizer management, litter and debris controls, bridge cleaning, maintenance and deck drainage, etc.

The conventional structural BMPs include (2):

✓ Permeable or Porous pavements
✓ Rain garden (or bioretention)

✓ Constructed wetlands

✓ Soil amendment and restoration

✓ Grassed Swales

✓ Filter Strips

✓ Sand Filters

✓ Infiltration Trenches

Porous or Permeable Pavements

Porous or pervious or permeable pavements represent a class of LID strategy that can address issues related to both stormwater quality and quantity (6). A porous pavement is composed of a permeable surface course underlain by a porous media reservoir. Plastic-reinforced turf grids, concrete pavers, porous concrete, and porous asphalt are some commonly used materials for the permeable surface course. An improvement in stormwater usually results from the combined effect of filtration and retention of pollutants in the surface layer, and water quality treatment in the porous media reservoir (7).

The popularity of porous pavements is steadily increasing with the evolving stormwater regulations and increasing sustainability-related concerns regarding water quality. It has been reported that, "with proper installation and maintenance, porous paving allows for infiltration of up to 80% of annual runoff volume" and "can remove between 65 and 85% of undissolved nutrients from runoff and up to

95% of sediment from runoff" (8). The reported benefits of permeable pavements are summarized in Fig. 8.2.

The use of porous asphalt pavements, pervious concrete, Permeable Interlocking Concrete Pavement (PICP), and Concrete Grid Paver (CGP) systems are included among US EPA's structural best practices for stormwater management (3).

Porous Asphalt Pavement

Porous asphalt has been in use since the 1970s to manage stormwater within the pavement structure. But, with the evolving stormwater regulations and increasing sustainability-related concerns regarding water quality, this technology is attracting increasing attention in recent years. They are especially well-suited for low-volume access roads and parking lots.

Porous asphalt is similar to conventional HMA except that the fine aggregates are removed which creates an air void matrix that allows water to drain through the pavement structure (see Fig. 8.3). A stone recharge bed stores the water until it eventually seeps into the natural underlying soil.

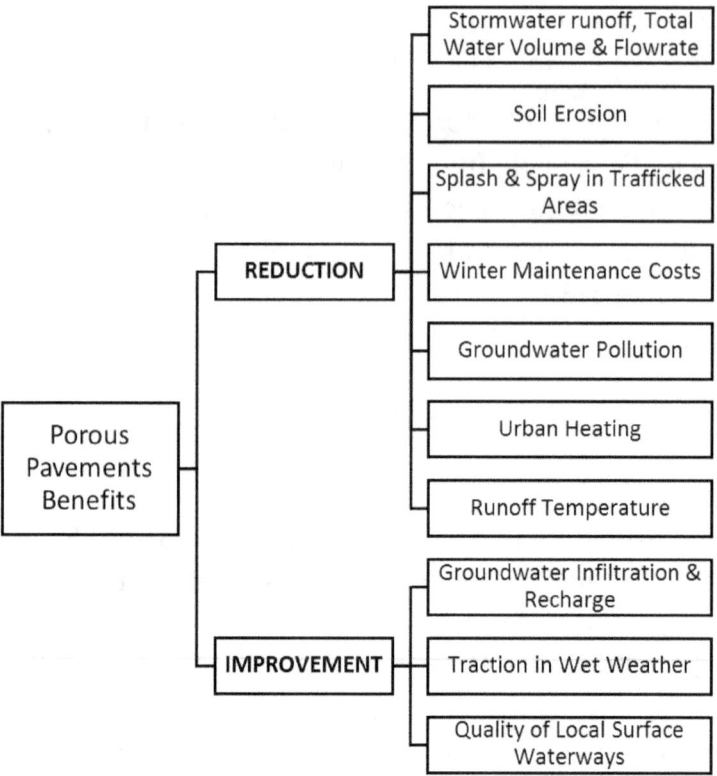

Fig. 8.2. Benefits of permeable pavements [*Source:* (6,9)]

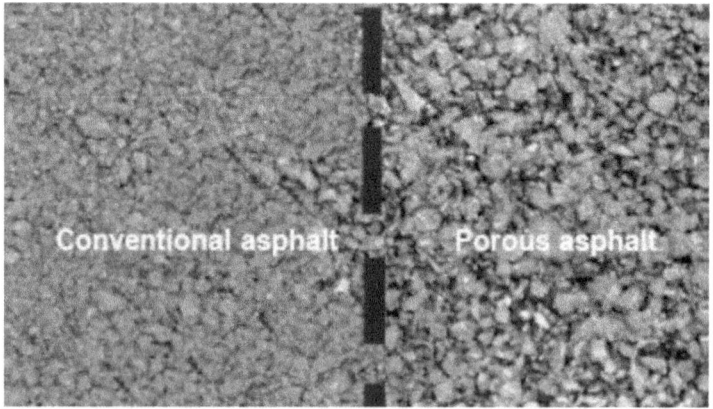

Fig. 8.3. Conventional and porous asphalt mixes [*Source:* Dauphin County Conservation District (8)]

A schematic of the porous asphalt system cross-section is shown in Fig. 8.4. A distinction is made in the literature between porous asphalt and asphalt open-graded friction course (OGFC) in terms of the system design and functionalities. While porous asphalt allows surface water to infiltrate into the porous media reservoir, OFGC permits only near-surface drainage with no surface infiltration into the base layer (7).

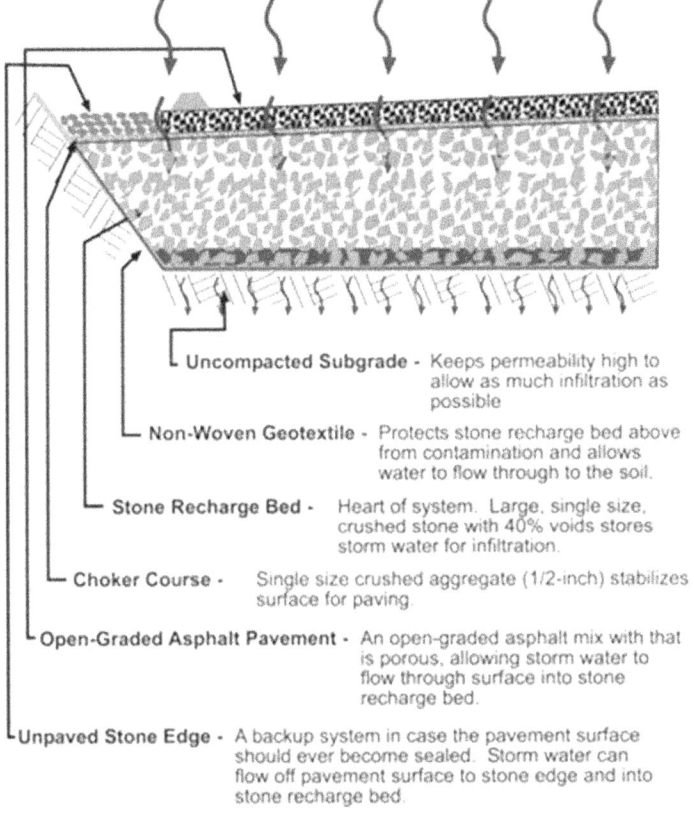

Fig. 8.4. A schematic of the porous asphalt system cross-section [*Source:* NAPA (10)]

Other benefits of using porous asphalt include reduction in noise, spraying and splashing from traffic, and improved skid resistance which is especially important in wet road conditions in cold climates (11,12).

There have been concerns regarding the migration of binder into the higher void spaces in early porous asphalt applications. This has been ameliorated with the use of additives (such as polymers) and additional binders which also enhance the characteristics of porous asphalt (13). Some studies have reported that porous asphalt is susceptible to clogging that cannot be cleaned by any combination of street sweeping or pressure washing and the only option is to remove and replace the surface layer every 8 years (14).

Pervious Concrete Pavement

Like traditional concrete, pervious concrete (also known as gap-graded or enhanced porosity concrete) is also made up of Portland cement, aggregate, water and optional additives, but it is gap-graded and contains very little or no fine sand, resulting in a coarse aggregate texture. The pervious concrete system consists of a pervious concrete layer on top of a coarse gravel retention layer for stormwater storage.

Pervious concrete has been commonly used in parking lots with a typical cross-section as follows: 4-6 inches of a pervious concrete layer, a permeable base with a thickness up to 18 inches, and a permeable subgrade. The hardened pervious concrete has about 15-25% air voids, typically. A typical cross-section of a pervious concrete pavement is shown in Fig. 8.5. In building pervious concrete in freeze-thaw areas, careful attention must be paid to control the degree of saturation (15).

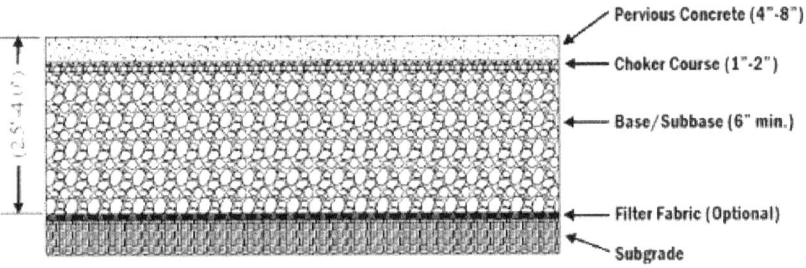

Fig. 8.5. A typical cross-section of pervious concrete pavement [*Source:* US EPA (16)]

Other benefits of using pervious concrete include enhancing pavement skid resistance during rainy days by rapid drainage of water, reducing pavement noise, and minimizing the heat island effect in large cities, and preserving the native ecosystems. Fig. 8.6 displays a pervious concrete specimen prepared in the laboratory showing a high void connectivity along sides.

Some reported disadvantages associated with the use of pervious concrete include clogging of voids, the requirement of a bi-yearly maintenance plan to unclog the voids and restore permeability, and the potential for groundwater contamination depending on the soil conditions. In addition, pervious concrete design in the US with single-sized coarse aggregate has resulted in high permeability, but not adequate strength and durability resulting in failures at an early stage of pavement life (15,17).

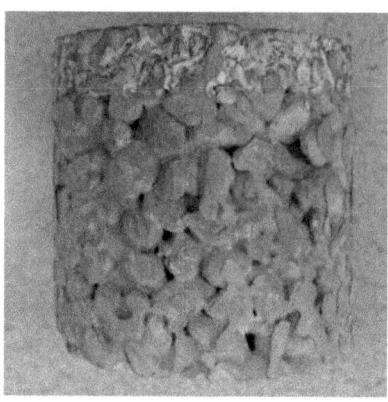

Fig. 8.6. A side view of a pervious concrete specimen showing high void connectivity along sides [*Source*: (15)]

Permeable Concrete Block Pavement

Concrete pavers, or porous paver blocks, or Permeable Interlocking Concrete Pavements (PICPs) allow water to drain through the openings created by the shape of the interlocking pavers (18). They can be considered as infiltration trenches with concrete pavers on top to support pedestrian movements and vehicular traffic. An open-graded base is typically used under PICPs because of the storage and treatment benefits. The captured runoff is stored in this high-void aggregate base detention reservoir until it percolates into the underlying subgrade or alternatively routed through a perforated underdrain system into a conventional stormwater conveyance (14). A typical schematic of the PICP schematic is shown in Fig. 8.7.

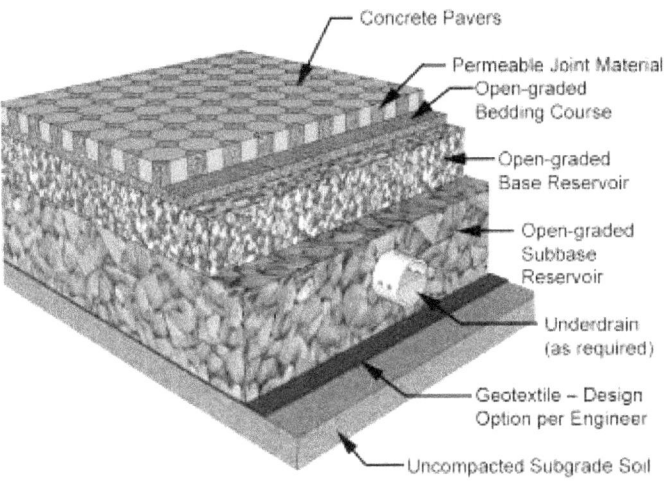

Fig. 8.7. Design schematic of Permeable Interlocking Concrete Pavement (PICP) [*Source:* ICPI (19)]

The four categories of PICPs include (20):

✓ Interlocking shapes with openings

✓ Enlarged permeable joint

✓ Porous concrete units

✓ Permeable pavers and concrete grid pavers

PICPs are different from Concrete Grid Pavements (CGPs) which are concrete units with cells that typically contain topsoil and grass. CGPs infiltrate water at rates lower than PICP and are generally not designed with an open-graded aggregate detention reservoir for storage. They are more suitable for intermittently trafficked areas such as parking areas and emergency fire lanes (18).

PICP applications (see Fig. 8.8) include pedestrian walkways, sidewalks, driveways, parking lots, and low-volume roadways (18). Other types of permeable concrete pavers include Modular Interlocking Concrete Bricks with Internal Drainage Cells (MICBID), and Modular Interlocking Concrete Bricks with External Drainage Cells (MICBEC) which provide drainage through open cells (14). Like other porous pavement types, periodic surface maintenance is considered a must to minimize clogging in interlocking concrete units with sediment to ensure long-term performance.

Fig. 8.8. Pictures of permeable concrete block pavements [*Source:* ICPI (20)]

Useful Online Resources

Stormwater BMPs

US EPA's Stormwater Program

http://cfpub.epa.gov/npdes/home.cfm?program_id=6

US EPA's National Menu of Stormwater Best Management Practices (BMPs)

http://cfpub.epa.gov/npdes/stormwater/menuofbmps/

International Stormwater BMP Database

http://www.bmpdatabase.org/

Federally and State Sponsored Stormwater BMP Manuals

http://environment.transportation.org/environmental_issues /construct_maint_prac/compendium/manual/3_7.aspx

Low-Impact Development (LID) Center

http://www.lowimpactdevelopment.org/index.html

University of New Hampshire (UNH) Stormwater Center

http://www.unh.edu/erg/cstev/

Porous Pavements

Porous Asphalt Pavements

http://cfpub.epa.gov/npdes/stormwater/menuofbmps/index. cfm?action=browse&Rbutton=detail&bmp=135&minmeasur e=5

NRMCA's Pervious Concrete Website

http://www.perviouspavement.org/

North Carolina State University's (NCSU's) Permeable Pavement Research Website

http://www.bae.ncsu.edu/info/permeable-pavement/index.html

Interlocking Concrete Pavement Institute (ICPI)

http://www.icpi.org/

Concrete Block Permeable Paving

http://www.paving.org.uk/permeable.php

References

1. Yu SL. NCHRP Synthesis of Highway Practice 174: Storm Water Management for Transportation Facilities. Washington, D.C.: Transportaion Research Board; 1993.

2. NCHRP. Management of Runoff fromSurface TransportationFacilities—Synthesis andResearch Plan [Internet]. Washington, D.C.: Transportaion Research Board; 2001 [cited 2011 Feb 7]. Available from: http://onlinepubs.trb.org/onlinepubs/nchrp/nchrp_w37.pdf

3. US EPA. National Pollutant Discharge Elimination System (NPDES): Stormwater Program [Internet]. EPA Stormwater Program - Office of Wastewater

Management. 2011 Feb 7 [cited 2011 Feb 7];Available from: http://cfpub.epa.gov/npdes/home.cfm?program_id=6

4. Pennsylvania Department of Environmental Protection (PA DEP). Pennsylvania Stormwater Best Management PracticesManual. PA: Bureau of Watershed Management, PA DEP; 2006.

5. Strecker EW, Quigley MM, Urbonas BR, Jones JE, Clary JK. Determining Urban Storm Water BMP Effectiveness. J. Water Resour. Plng. and Mgmt. [Internet]. 2001 Jun [cited 2011 Feb 7];127(3):144-149. Available from: http://link.aip.org/link/?QWR/127/144/1

6. Ferguson BK. Porous pavements. Boca Raton, Florida: CRC Press; 2005. 604 p.

7. Briggs JF. Performance Assessment of Porous Asphalt for Stormwater Treatment [Internet] [M.S. Thesis]. [Durham, NH]: University of New Hampshire; 2006 [cited 2011 Feb 7]. 212 p.Available from: https://unh.edu/erg/cstev/pubs_specs_info/unhsc_briggs_thesis_12_06.pdf

8. Dauphin County Conservation District. Porous Asphalt: Best Management Practices Fact Sheet [Internet]. 2011 Feb 7 [cited 2011 Feb 7];Available from: http://www.dauphincd.org/swm/BMPfactsheets/Porous%20Asphalt%20fact%20sheet.pdf

9. University of Washington. The Greenroads Rating System [Internet]. [cited 2011 Mar 28];Available from: http://www.greenroads.us/

10. NAPA. Innovative Stormwater Management: Porous Asphalt Pavement with Subsurface Recharge Bed

[Internet]. [cited 2011 Apr 2];Available from:
http://www.hotmix.org/images/stories/porous_asphalt_
poster1.pdf

11. Moore L, Hicks R, Rogge D. Design, Construction, and
Maintenance Guidelines for Porous Asphalt Pavements.
Transportation Research Record [Internet]. 2001 [cited
2011 Feb 7];1778:91-99. Available from:
http://pubsindex.trb.org/view.aspx?id=717354

12. US EPA. Storm Water Technology Fact Sheet: Porous
Pavement [Internet]. Washington, D.C.: US EPA, Office
of Water; 1999 [cited 2011 Feb 7]. Available from:
http://www.epa.gov/npdes/pubs/porouspa.pdf

13. Hun-Dorris T. Advances in Porous Pavement.
Stormwater: The Journal for Surface Water Quality
Professionals [Internet]. 2005 [cited 2011 Feb 7];(March-
April). Available from:
http://www.stormh2o.com/march-april-2005/pavement-
materials-watershed.aspx

14. James W, Langsdorff H. The Use of Permeable Concrete
Block Pavement in Controlling Environmental Stressors
in Urban Areas [Internet]. Sun City, South Africa:
Document Transformation Technologies; 2003 [cited
2011 Feb 8]. Available from: http://www.environmental-
expert.com/Files/11067/articles/4871/027.pdf

15. Schaefer VR, Wang K, Suleiman MT, Kevern JT. Mix
Design Development for Pervious Concrete In Cold
Weather Climates [Internet]. Ames, IA: National
Concrete Pavement Technology Center, Iowa State
University; 2006 [cited 2011 Feb 7]. Available from:
http://www.intrans.iastate.edu/reports/mix_design_per
vious.pdf

16. US EPA. Pervious Concrete Pavement [Internet]. EPA - Stormwater Menu of Best Management Practices (BMPs). 2009 Sep [cited 2011 Feb 8];Available from: http://cfpub.epa.gov/npdes/stormwater/menuofbmps/index.cfm?action=browse&Rbutton=detail&bmp=137&minmeasure=5

17. US EPA VR. Post-Construction Storm Water Management in NewDevelopment and Redevelopment [Internet]. Ames, IA: US Environmental Protection Agency (EPA); 2004 [cited 2011 Feb 7]. Available from: http://cfpub.epa.gov/npdes/stormwater/menuofbmps/pdf/final/sec-5_new.pdf

18. US EPA. Permeable Interlocking Concrete Pavement [Internet]. EPA - Stormwater Menu of Best Management Practices (BMPs). 2009 Sep [cited 2011 Feb 8];Available from: http://cfpub.epa.gov/npdes/stormwater/menuofbmps/index.cfm?action=browse&Rbutton=detail&bmp=136&minmeasure=5

19. ICPI. Permeable Interlocking Concrete Pavement: Schools and Universities Fact Sheet [Internet]. 2008 [cited 2011 Feb 7];Available from: http://www.icpi.org/sites/default/files/SchoolsUniversitiesPICP-20.pdf

20. Smith DR. Permeable Interlocking Concrete Pavements: Selection, Design, Construction & Maintenance [Internet]. Third Edition. Interlocking Concrete Pavement Institute (ICPI); 2006 [cited 2011 Feb 7]. 62 p.Available from: http://www.wgpaver.com/files/PDF/PICP_Manual_3rd_edition.pdf

K. Gopalakrishnan

Chapter 9:
Cool Pavements

According to the US EPA (1), cool pavements refers to "paving materials that reflect more solar energy, enhance water evaporation, or have been otherwise modified to remain cooler than conventional pavements". A significant percentage of a community's land is covered by pavements which can absorb and store much of Sun's energy contributing to urban heat island effect, more than roof and vegetated surfaces. Therefore, cool pavement technologies are an important element to consider in Urban Heat Island (UHI) mitigation.

Pavements and Urban Heat Island

There are two primary mechanisms through which pavements contribute to UHI (2): (1) the materials properties of conventional paving materials causes them to absorb and store a large amount of heat than vegetated land cover; and (2) in comparison to vegetated land cover, the impervious surface of most pavements reduces cooling due to evaporation. As a result, cool pavement technologies have primarily focused on (1) increasing the solar reflectance of pavements through use of lighter-colored paving materials, and (2) improving cooling through the use of porous paving materials. The heat-related processes that can affect a pavement structure are illustrated in Fig. 9.1.

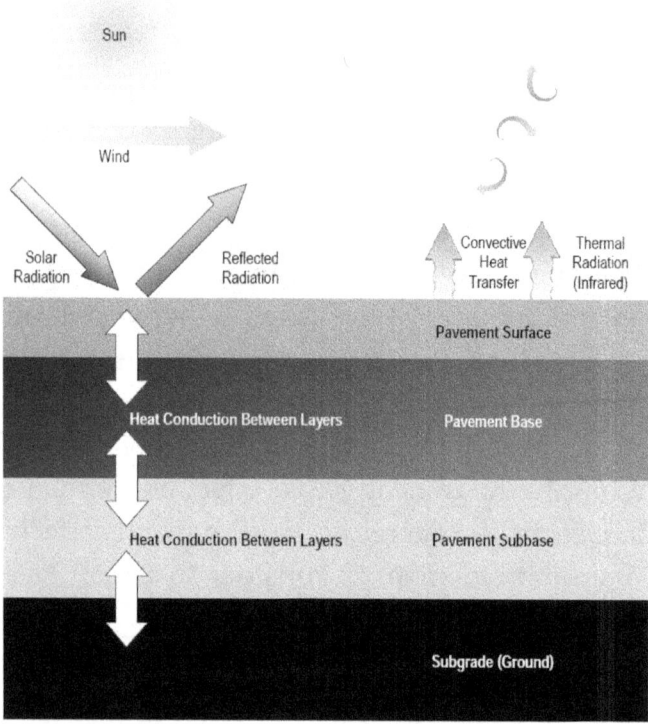

Fig. 9.1. Pavement heat-related characteristics and processes [*Source:* (2)]

Albedo or Solar Reflectance

Pavement temperatures are significantly influenced by solar energy, solar reflectance, material heat capacities, surface roughness, heat transfer rates, thermal emittance, and permeability. Existing literature on cool pavements have mainly focused on pavement solar reflectance (albedo) which is the primary determinant of a material's maximum surface temperature. The premise behind this approach is that if pavements are designed to reduce the absorption of solar energy (i.e., higher solar reflectance), consequently less heat will be radiated to the surrounding environment.

Albedo or solar reflectance is a measure of solar energy reflected by a surface. While albedo is typically measured on a scale of 0 to 1, it is also sometimes reported as a percentage. Albedos of conventional asphalt and concrete pavements vary with age due to weathering and accumulation of dirt and can range from 5 to 40%. As seen in Fig. 9.2, conventional asphalt pavements tend to lighten (higher solar reflectance) with time since the asphalt binder oxidizes and more aggregate is exposed through wear. On the other hand, conventional concrete pavements tend to darken (lower solar reflectance) with time due to vehicular traffic and other factors.

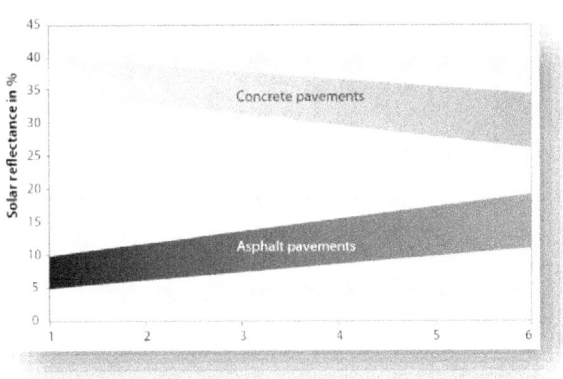

Fig. 9.2. Typical albedos of conventional pavements over time [*Source:* US EPA (3)]

201

Apart from pavement age and aggregate color, other factors such as wind, shadowing trees, and even vehicle shadows contribute to pavement solar reflectivity. In general, light-colored pavements with high albedos can significantly reduce the pavement surface temperatures and contribute to UHI mitigation (see Fig. 9.3).

Fig. 9.3. A picture showing how lighter colored pavements are substantially cooler than nearby darker roads [*Source:* K. Kaloush (4)]

Thermal Emittance

Another important material property which plays a role in determining a material's contribution to UHI is thermal emittance. Albedo affects a how a pavement heats up (maximum surface temperatures) while emittance affects how a pavement cools down (minimum surface temperatures) (5). However, most research studies have focused on modifying the albedo of pavement materials since most pavement materials inherently have high emittance values, thus limiting the number of options available to modify this property to achieve cool pavements (6).

Permeability

Permeable or porous pavements (porous asphalt, pervious concrete, permeable pavers, and grid pavements) discussed previously are emerging as potential cool pavements owing to their ability to allow air, water, and water vapor into the pavement voids, keeping the material cool when moist. Similar to evaporative cooling from vegetated land cover, these pavements can lower temperatures by evaporating the moisture within the pavement structure as the surface heats up drawing heat out of the pavement.

Other factors that are considered to influence pavement's heat transfer characteristics and processes include convention, thermal conductivity of paving materials, heat capacity of pavements, pavement thickness, and urban geometry (1).

Potential Cool Pavement Types, Benefits and Costs

According to US EPA's Compendium of Strategies for Reducing UHI (3), the following potential cool pavement types exist:

- Conventional asphalt pavements modified with high albedo materials or treated after installation to raise reflectance
- Conventional concrete pavements
- Other reflective pavements
  - Resin based pavements
  - Colored asphalt
  - Colored concrete
- Nonvegetated permeable pavements
  - Porous asphalt
  - Rubberized asphalt
  - Pervious concrete
  - Brick or block pavers
- Vegetated permeable pavements
- Chip seals
- Whitetopping
- Ultra-thin whitetopping
- Microsurfacing

The reported potential benefits of cool pavements are as follows (3,7):

✓ Reduced UHI effect

✓ Reduced energy use and GHG emissions

✓ Improved water quality and reduced stormwater runoff (using permeable pavement)

✓ Increased pavement life/durability and waste reduction

✓ Reduced pavement maintenance costs

✓ Enhanced nighttime illumination

✓ Comfort improvements

✓ Improved driving safety (using permeable pavement)

✓ Noise reduction

A number of factors will influence the costs of cool pavements such as the region in which the cool pavement needs to be constructed, local climate, contractor, time of year, accessibility of the site, underlying soils, project size, expected traffic, and the desired pavement service life. Based on available sources, the US EPA Compendium of Strategies for Reducing UHI (3) provides a summary of a range of costs associated with both conventional and cool pavements (see Fig. 9.4). It is noted that these costs do not reflect life-cycle costs and are provided as estimates for initial construction or performing maintenance.

Basic Pavement Types	Example Cool Approaches	Approximate Installed Cost, $/square foot*	Estimated Service Life, Years
New Construction			
Asphalt (conventional)	Hot mix asphalt with light aggregate, if locally available	$0.10–$1.50	7–20
Concrete (conventional)	Portland cement, plain-jointed	$0.30–$4.50	15–35
Nonvegetated permeable pavement	Porous asphalt	$2.00–$2.50	7–10
	Pervious concrete	$5.00–$6.25	15–20
	Paving blocks	$5.00–$10.00	> 20
Vegetated permeable pavement	Grass/gravel pavers	$1.50–$5.75	> 10
Maintenance			
Surface applications	Chip seals with light aggregate, if locally available	$0.10–$0.15	2–8
	Microsurfacing	$0.35–$0.65	7–10
	Ultra-thin whitetopping	$1.50–$6.50	10–15

* Some technologies, such as permeable options, may reduce the need for other infrastructure, such as stormwater drains, thus lowering a project's overall expenses. Those savings, however, are not reflected in this table. (1 square foot = 0.09 m²)

Fig. 9.4. Comparative costs for conventional and cool pavement types [*Source:* US EPA (3)]

Useful Online Resources

US EPA's Heat Island Effect Website

http://www.epa.gov/heatisland/index.htm

Lawrence Berkeley National Laboratory (LBNL) Heat Island Group

http://eetd.lbl.gov/HeatIsland/

US EPA's Cool Pavements Website

http://www.epa.gov/heatisland/mitigation/pavements.htm

Arizona State University (ASU) National Center of
Excellence on SMART Innovations

http://asusmart.com/

References

1. US EPA. Cool Pavements [Internet]. 2009 Feb 9 [cited
    2011 Feb 24];Available from:
    http://www.epa.gov/heatisld/mitigation/pavements.ht
    m

2. Cambridge SystematicsInc. Cool Pavement Report -
    EPA Cool Pavements Study - Task 5 [Internet].
    Washington D.C.: US Environmental Protection
    Agency (EPA); 2005 [cited 2011 Feb 24]. Available from:
    http://www.epa.gov/heatisld/resources/pdf/CoolPavem
    entReport_Former%20Guide_complete.pdf

3. US EPA. Reducing Urban Heat Islands: Compendium
    of Strategies - Cool Pavements [Internet]. Washington
    D.C.: US Environmental Protection Agency (EPA); 2006
    [cited 2011 Feb 24]. Available from:
    http://www.epa.gov/heatisld/resources/pdf/CoolPavesC
    ompendium.pdf

4. Kaloush K. Pavements and the Urban Heat Island Effect
    [Internet]. 2010 Jan 28 [cited 2011 Mar 18];Available
    from:
    http://www.epa.gov/heatisld/resources/pdf/2ScientificO
    verview-Pavements-HeatIslands-Kaloush-ASU.pdf

5.   National Center of Excellence on SMART Innovations at
      Arizona State University. What Factors Influence
      Elevated Pavement Temperatures Most During Day
      and Night? Case Study 1(1). 2007;

6.   Levinson R, Akbari H, Konopacki S, Bretz S. Inclusion
      of Cool Roofsin Nonresidential Title 24Prescriptive
      Requirements. Berkeley, CA: Lawrence Berkeley
      National Laboratory (LBNL); 2002.

7.   Pavement Interactive. Cool Pavement: General -
      Pavement Interactive [Internet]. 2010 Apr [cited 2011
      Feb 24];Available from:
      http://pavementinteractive.org/index.php?title=Cool_Pa
      vement:General

Chapter 10:
Quiet Pavements

Highway Traffic and Construction Noise

Noise from highway traffic and construction causes annoyance to humans contributing to decreased quality of life as well as negatively influence the well-being of other ecosystems (1).One study indicated that 30% of people living in Europe are exposed to traffic noise exceeding the World Health Organization (WHO) recommended acceptable levels and a significant part of the economy of those countries are affected by noise impact (2,3). Fig. 10.1 shows typical sound levels that people experience (4). Traffic noise measurements are typically reported in units of dBA (rather than dB), which is a weighted measure of how the human ear perceives and responds to sound (5).

In the US, the construction of noise barrier walls has been the primary strategy for mitigating noise from highway operations. However, these barrier walls are not only expensive to build and maintain, but their effectiveness is limited, often to less than one-fourth of a mile from the roadway (6). According to the FHWA, between 1970 and 1998, 1,630 miles of sound barriers were built in the US at a cost of $1.9 billion (7). This has prompted research in the direction of attacking the noise problem at its source. Among the different source control strategies, the use of quiet pavement technologies, which have a longer history in Europe, are considered as promising alternatives in addressing highway noise problems (6).

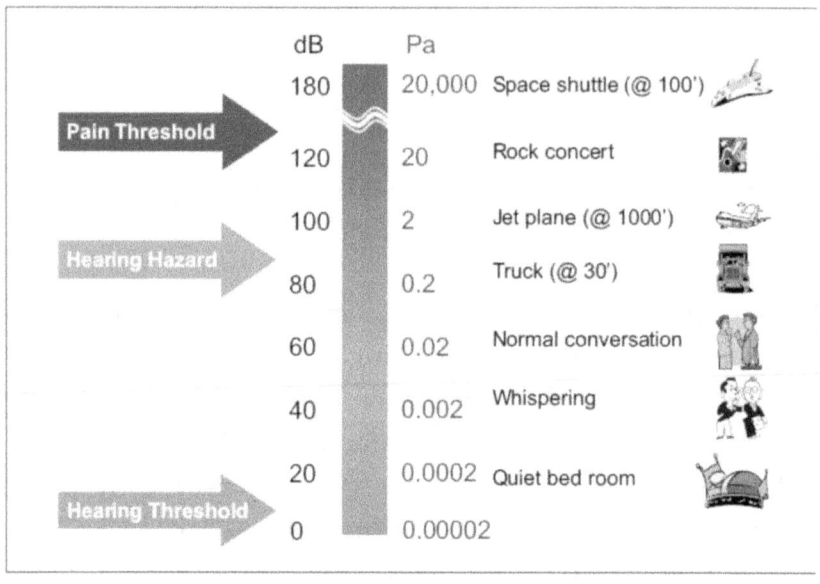

Fig. 10.1. Sound levels and typical experiences [*Source:* (4)]

Recent interest in quiet pavements has been driven by cost considerations and an increasing public awareness regarding the effect of pavement selection on resultant traffic noise levels. Especially in the recent context of constructing sustainable and environmental friendly highways, quiet pavement technologies are receiving a lot of attention although the idea of designing and building quieter pavements is not new.

Tire-Pavement Interaction and Pavement Surface Characteristics

A majority of highway noise originates at the tire/pavement interface at automobile speeds greater than 30 mph. Several studies have indicated that the use of quieter tires or quieter pavements could substantially reduce traffic-generated noise (1). Among the many factors that play a role in the generation of sound due to tire-pavement interaction, pavement surface characteristics have a significant influence on traffic noise level variations, assuming that all other factors are constant.

The noise reduction mechanisms by the pavement itself depend on the pavement surface texture (i.e., microtexture, macrotexture, etc.), porosity, and the relative stiffness of the tire and the pavement, in decreasing order of importance. A schematic illustration of the classification of textures and the associated pavement surface characteristics is shown in Fig. 10.2 (8)

The NCHRP project 01-44 (9) evaluated potential tire-pavement noise measurement procedures and new pavements incorporating noise-mitigating features, and identified design and construction features associated with different noise levels and the research results were reported in NCHRP Report 630 (10). A significant outcome of this research study was the development of a draft On-Board Sound Intensity (OBSI) test procedure for measuring at-the-source tire-pavement noise. The OBSI measurements can help assess which pavements are louder than others.

According to the FHWA's *Little Book on Quieter Pavements* (11), both asphalt and concrete pavements can be engineered with noise reduction properties to achieve the same level of cost-effectiveness, durability, and safety.

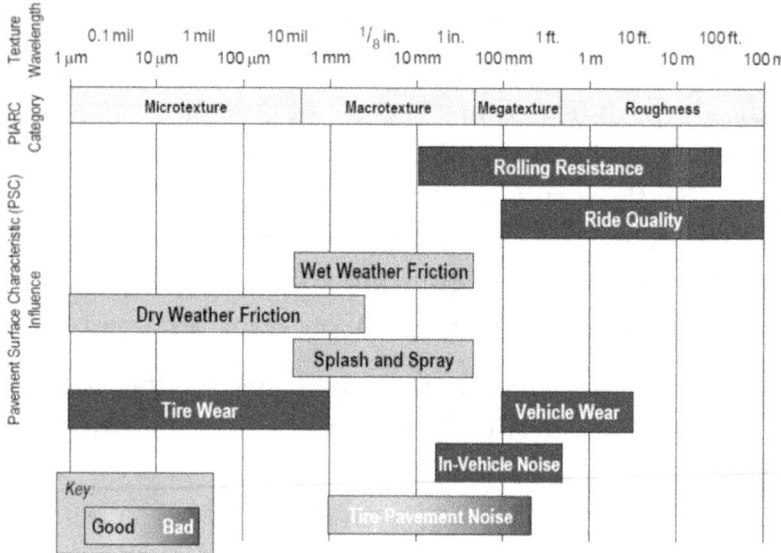

Fig. 10.2. Pavement surface texture classifications and associated characteristics [*Source:* CP Tech Center (8)]

Quiet Asphalt Pavements

The major noise-reducing asphalt pavement technologies that are used by several European countries, US and Japan include open-graded (porous) asphalt, rubberized asphalt, and gap-graded asphalt or Stone Mastic Asphalt (SMA). Illustrations and photographs of open-graded asphalt and SMA are displayed in Fig. 10.3. The specific quiet asphalt mixture types in use are listed below:

- ✓ Open-Graded Asphalt Concrete (OGAC)/ Open-Graded Asphalt Friction (wearing) Course (OGFC)

- ✓ Rubberized OGAC/Asphalt Rubber OGFC

- ✓ Rubberized Gap-Graded Asphalt Concrete (RAC-G)

- ✓ Stone Matrix Asphalt (SMA)

- ✓ Single- and Double-layer Highly Porous Asphalt Mixes

(a)

(b)

Fig. 10.3. Quieter asphalt pavement technologies: (a) schematic and photograph of open-graded (porous) asphalt; (b) schematic and photograph of gap-graded SMA [*Source:* FHWA (11)]

The open-graded (porous) mixtures have lesser fine aggregate material and air void contents higher than 15% in general. The interconnected air voids help in both reducing the generated noise generation as well as absorbing the generated noise in the air void structure (12). However, there are concerns regarding the durability of these pavements and the loss of noise reducing properties within a short time.

Due to high winter maintenance costs associated with the prevention of black ice formation on highly porous pavements, they are not used by some European countries in snow and ice regions. Also, highly porous mixes are not recommended to be placed in urban areas as they tend to clog under slow speed (less than 45 mph) in urban areas.

SMA is a gap-graded mixture that "leaves out" intermediate sized material and it works by creating a negative texturing texture and/or using smaller maximum aggregate sizes (11). There is a concern that since mixtures like SMA are not primarily geared towards noise reduction, it may be possible that they may lose their noise reduction capabilities more quickly than those pavements specifically engineered to mitigate noise (13). Also, the skid resistance of SMA is an issue, especially during the early life of the pavement (2).

One study (14) evaluated tire/pavement noise data from more than 60 in-service highway asphalt pavements and examined the effects of surface microtexture, aggregate gradation, permeability, and air-voids content in the mix on tire/pavement noise levels. Data were collected on OGAC, rubberized OGAC, RAC-G, as well as regular dense graded asphalt concrete mixtures. The findings confirmed that open-graded pavements are quieter than denser pavements.

Reduced noise levels were correlated to higher air void contents while higher noise levels were correlated to higher levels of macrotexture. It was also reported that the open-graded mixes tend to lose their noise mitigation properties with time due to clogging as well as due to pavement surface distresses. Based on the study results, it was recommended that best approach currently available to noise reduction is to use thin-surfaced open-graded mixes with smaller nominal maximum aggregate sizes (14).

Quiet Concrete Pavements

Quieter concrete pavements are achieved through the use of open-graded PCC as well as through various PCC surface texturing methods listed below:

✓ Drag textures (including artificial turf and burlap drag)

✓ Longitudinal tining

✓ Transverse tining

✓ Diamond grinding (also known as whisper grinding)

Among these, the conventional texturing (drag and tining) is performed while concrete is still in a plastic state while diamond grinding is performed on hardened concrete. Artificial turf drag surfaces are created by dragging a piece of artificial turf across the pavement surface and burlap drag texturing is created by dragging moistened, coarse burlap across the surface (8).

Transversely tined textures (most commonly used texturing methods in the US) are created by moving the tining device across the pavement width whereas longitudinally tined textures are created by moving the tining device longitudinally along the direction of paving (8). Diamond grinding, which has been traditionally used for PCC surface restoration, employs closely spaced diamond saw blades to remove a thin layer of hardened concrete pavement (8).

Fig. 10.4 illustrates the normalized distributions of noise levels that are possible for each nominal texture type based on data collected from nearly 1,500 concrete pavement textures worldwide (15). It has been reported that concrete pavement texturing life depends on traffic and the presence of studded tires. While the durability of carpet drag surfaces is not yet well understood, tining has been shown to last from anywhere between 3 to 6 years or more. A longer texturing life (10 to 17 years) is diamond ground PCC surfaces (13). A recent study indicated the possibility of reducing tire-pavement noise through improved joint construction techniques in addition to surface texture enhancements.

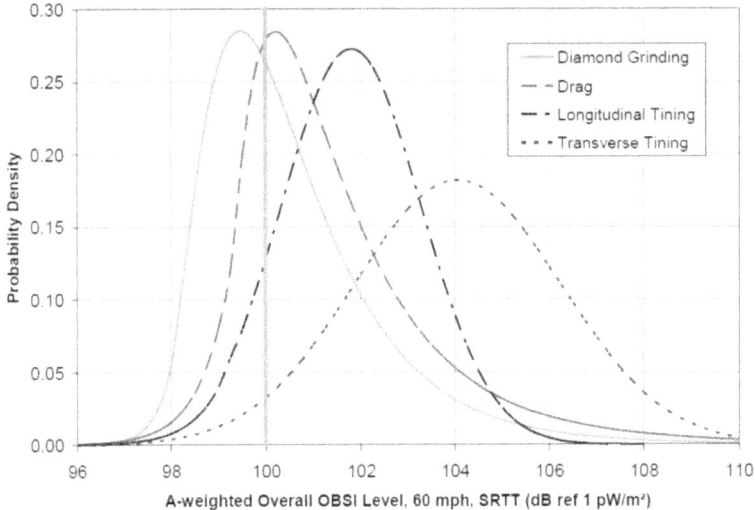

Fig. 10.4. Ranges of noise levels measured on concrete pavement surfaces [*Source:* (15)]

Other quieter concrete pavement solutions that are being researched include exposed aggregate surfaced concrete pavements, pervious concrete pavements, stamping, brushing, other new texturing techniques, sprinkle treatment, and shot peening. Exposed aggregate concrete pavements, which are commonly constructed using a two-layer "wet on wet" paving process, are more prevalent in European countries and has not been routinely used in the US. Pictures of the most commonly used quieter concrete pavement technologies are shown in Fig. 10.5.

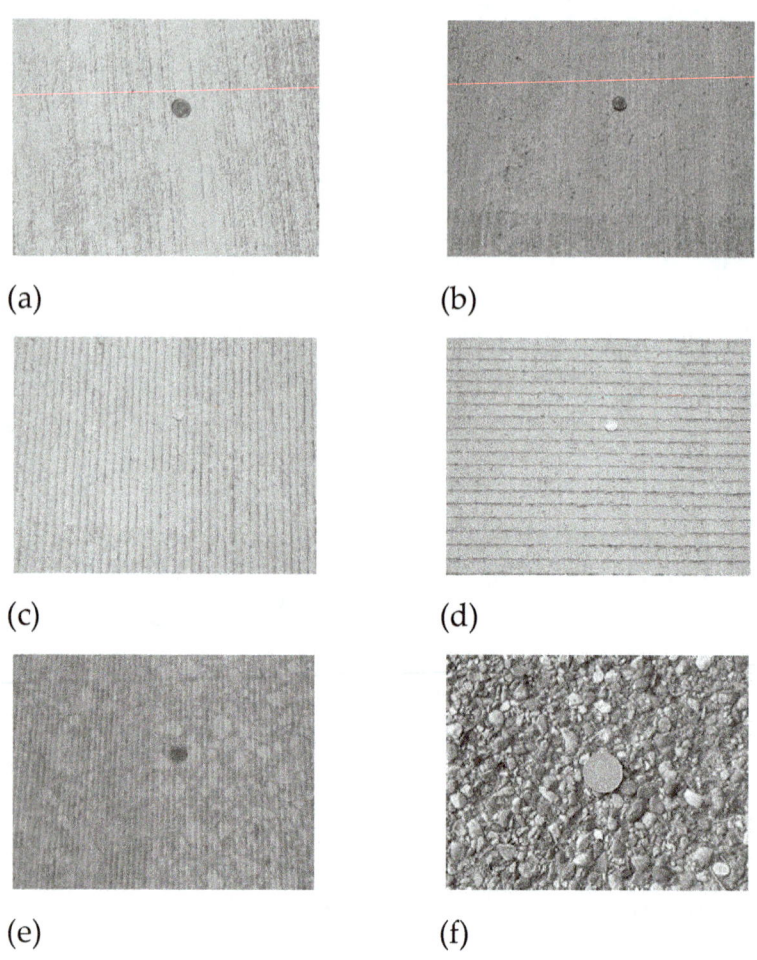

Fig. 10.5. Quieter concrete pavement alternatives: (a) artificial turf drag texture, (b) burlap drag texture, (c) longitudinal tining, (d) transverse tining, (e) diamond ground surface, and (f) finished exposed aggregate surface [*Source*: CP Tech Center (8)]

Useful Online Resources

FHWA Highway Traffic Noise Website

http://www.fhwa.dot.gov/environment/noise/

Quiet Pavement Systems in Europe: FHWA International Scan Tour

http://international.fhwa.dot.gov/quiet_pav/

SILVIA: Sustainable Road Surfaces for Traffic Noise Control – A Project Initiated by the Forum of European National Highway Research Laboratories (FEHRL)

http://www.trl.co.uk/silvia/

APA's Quiet Asphalt Pavement Website

http://www.quietpavement.com/

Concrete Pavement Surface Characteristics Program (CPSCP)

http://www.surfacecharacteristics.com/

References

1. Bernhard R, Wayson RL. An introduction to tire/pavement noise. West Lafayette, IN: Institute for Safe, Quiet and Durable Highways. Purdue University; 2005.

2. Ahammed MA. Safe, Quiet and Durable Pavement Surfaces [Internet] [Ph.D. Dissertation]. [Waterloo, Ontario, Canada]: University of Waterloo; 2009 [cited 2011 Feb 5]. 292 p.Available from: http://uwspace.uwaterloo.ca/bitstream/10012/4290/1/Mo

hammad%20Alauddin%20Ahammed%20-
%20Ph.D.%20Thesis.pdf

3. FEHRL. Guidance Manual for the Implementation of
    Low-Noise Road Surfaces [Internet]. Brussels, Belgium:
    FEHRL; 2006 [cited 2011 Feb 5]. Available from:
    http://www.trl.co.uk/silvia/Silvia/pdf/silvia_guidance_
    manual.pdf

4. Rasmussen RO, Sahoney R, Weigand P, Harrington D.
    The Language of Noise and Quieter Pavements: Noise
    101 Tech Brief [Internet]. 2010 Oct [cited 2011 Feb
    5];Available from:
    http://www.cptechcenter.org/publications/Noise_101_te
    ch_brief.pdf

5. FHWA. Procedures for Abatement of Highway Traffic
    Noise and Construction Noise: 23 CFR 772 [Internet].
    FHWA, U.S. Department of Transportation. 2001 [cited
    2011 Feb 5];Available from:
    http://www.fhwa.dot.gov/hep/23cfr772.htm

6. Gibbs D, Iwasaki R, Bernhard R, Bledsoe J, Carlson D, et
    al. Quiet Pavement Systems in Europe [Internet].
    Washington D.C.: Federal Highway Administration
    (FHWA); 2005 [cited 2011 Feb 3]. Available from:
    http://international.fhwa.dot.gov/quiet_pav/pl05011.pdf

7. Kandhal PS. Asphalt Pavements Mitigate
    Tire/Pavement Noise. HMAT Magazine [Internet]. 2004
    [cited 2011 Feb 4];(March/April):22-31. Available from:
    http://www.flexiblepavements.org/images/kandhalnois
    eart.pdf

8. Cackler ET, Ferragut T, Harrington D, Rasmussen RO,
    Weigand P, et al. Evaluation of U.S. and European

Concrete Pavement Noise Reduction Methods [Internet]. Ames, IA: National Concrete Pavement Technology Center, Iowa State University; 2006 [cited 2011 Feb 5]. Available from: http://www.intrans.iastate.edu/reports/surface_characteristics_i.pdf

9. NCHRP. NCHRP 01-44: Measuring Tire-Pavement Noise at the Surface [Internet]. 2011 Feb 4 [cited 2011 Feb 4];Available from: http://144.171.11.40/cmsfeed/TRBNetProjectDisplay.asp ?ProjectID=230

10. Donavan PR, Lodico DM. Measuring Tire-Pavement Noise at the Surface [Internet]. Washington, D.C.: Transportaion Research Board; 2009 [cited 2011 Feb 4]. Available from: http://onlinepubs.trb.org/onlinepubs/nchrp/nchrp_rpt_630.pdf

11. Rasmussen RO, Bernhard R, Sandberg U, Mun EP. The Little Book of Quieter Pavements [Internet]. Washington D.C.: Federal Highway Administration (FHWA); 2007. Available from: http://www.tcpsc.com/LittleBookQuieterPavements.pdf

12. Sandberg U, Ejsmont JA. Tyre/Road Noise Reference Book. Handelsbolag, Sweden: Informex; 2002.

13. Muench ST, Anderson JL, Hatfield JP, Koester JR, Soderlund M. Greenroads Rating System v1.5. Seattle, WA: University of Washington;

14. Kohler E, Motumah L, Harvey J. Tire/Pavement Noise and Other Surface Characteristics from Highway Pavements in California. In: CONINFRA 2009 -

Transportation Infrastructure Congress. Sao Paulo, Brasil: 2009.

15. Rasmussen RO, Garber SI, Fick GJ, Ferragut TR, Weigand P. How to Reduce Tire-Pavement Noise: Interim Better Practices for Constructing and Texturing Concrete Pavement Surfaces [Internet]. Ames, IA: National Concrete Pavement Technology Center, Iowa State University; 2008 [cited 2011 Feb 5]. Available from: http://www.intrans.iastate.edu/reports/surface_characte ristics_i.pdf

Chapter 11:
Green Rating Systems for Highways and Pavements

What exactly is a "green" pavement? How do we assess the sustainability aspects of a highway project? The answers to these questions are explored in this chapter by providing an up to date discussion on the existing and recent roadway sustainability assessment and rating/certification systems, which are referred to as "green" rating systems.

The Leadership in Energy and Environmental Design (LEED™) green building rating system developed by the U.S. Green Building Council's (USGBC) is the nationally recognized standard for measuring building sustainability. Almost all of the highway sustainability rating systems developed to date are modeled after the LEED™ rating system. It is recognized that a single scale rating system may not be able to cater to every region of the US and any new green highway rating system will also need to be flexible to change (1).

According to the FHWA, measuring sustainability of highway projects can have several benefits (2):

✓ Help establish standard and qualitative measures for sustainability that will enable agencies and program managers to track and assess progress

✓ Encourage broad participation in establishing highway sustainability goals

✓ Help users to better evaluate sustainability tradeoffs resulting in better and more rational decision-making

✓ Help users effectively meet emerging regulations such as those related to climate change and GHG emissions

✓ Reward excellence by conferring market recognition for sustainability efforts and application of sustainability best practices thereby increasing the general awareness of sustainability efforts and rewarding those who participate in the effort

✓ Effectively communicate sustainability goals and benefits to stakeholders with increased transparency

Currently, there is no requirement in any of the green rating systems (including the FHWA's rating system) to have a transportation project review prior to construction, although designers and contractors may choose to do so. In these systems, typically more points are awarded for sustainability strategies that will have greater positive impacts on energy efficiency and CO_2 emissions. An important point to keep in mind is that these green rating systems are constantly under development by design. Although the information presented here reflects the latest status with respect to each of them and are subject to revision, it is expected that the overall framework may not change significantly in a short time.

LEED™

LEED™ (Leadership in Energy and Environmental Design) is a voluntary, third-party green building certification system introduced by the United States Green Building Council (USGBC®) in 1998 to facilitate global adoption of sustainable green building and development practices. Over the years, it has undergone several revisions and has evolved into a nationally accepted benchmark for the design, construction and operation of high-performance green buildings of various types in the US. Some of the worldwide countries using the LEED™ system include Canada, Brazil, Mexico, and India. The USGBC, Inc. is a non-profit organization with currently more than 18,000 member companies and organizations representing the entire building industry (3).

An open, consensus-based process led by the USGBC's LEED™ committees is used in the on-going development and revision of the different LEED™ rating systems. Currently, they are applicable to six different markets: New Construction (LEED™-NC), Existing Buildings (LEED™-EB), Commercial Interiors (LEED™-CI), Core & Shell (LEED™-CS), Homes (LEED™-H), and Neighborhood Development (LEED™-ND, in pilot phase). Some of these markets include a suite of LEED™ rating systems such as that for New Construction (some of which are still pilot programs): LEED™ for Multiple Buildings/Campuses, LEED™ for Schools, LEED™ for Healthcare, LEED™ for Retail, and LEED™ for Laboratories.

A whole-building approach to green design and sustainability is promoted by LEED™ by recognizing performance in the following seven key areas:

✓ Sustainable Sites (SS)

✓ Water Efficiency (WE)

✓ Energy and Atmosphere (EA)

✓ Materials and Resources (MR)

✓ Indoor Environmental Quality (EQ)

✓ Innovation in Design (ID)

✓ Regional Priority (RP)

Each of these categories is broken down into prerequisites and credits as shown in Fig. 11.1. Prerequisites are mandatory for all projects and receive no points at all whereas credits are optional and have a range of possible points. The LEED™ standard does not completely prescribe how savings should be realized. As a result, many see its flexibility in providing a range of opportunities to get the LEED™ rating as part of the benefit of the LEED™ (4).

To obtain LEED™ certification, the building project team should submit documents showing evidence of meeting the LEED™ compliance requirements. Upon satisfactory review and compliance verification, the USGBC issues the LEED™ certification for a certain fee. Building projects can qualify for four certification levels in LEED™:

✓ *Certified* (40 – 49 points)

✓ *Silver* (50 – 59 points)

✓ *Gold* (60 – 79 points)

✓ *Platinum* (80 points and above)

On April 27, 2009, the USGBC launched the latest version of LEED™: LEED™ v3 (Version 3) (5) which incorporates the latest technology in building science and addresses the most urgent priorities like energy use and CO_2 emissions. As of December 2, 2010, more than 6,000 non-residential buildings have received LEED™ certification. A report published in 2003, based on a detailed review of 60 LEED™ rated buildings, found that green buildings are on an average 25-30% more energy efficient compared to conventional buildings and are characterized by reduced electricity purchases and reduced peak energy demand (6). According to the Urban Green Council, the New York Chapter of USGBC, a typical green building saves 30% in energy, 35% in GHG emissions, and up to 50% in water usage (7).

The USGBC established the Green Building Certification Institute (GBCI) as an independent, third-party organization that offers the highest-level validation that green building performance (i.e., LEEDTM building certifications) and green building practices (i.e., LEEDTM professional credentials) have met specific, rigorous criteria. The GBCI manages the LEEDTM professional credentials through a series of exams that allow individuals to become accredited for their knowledge of the LEEDTM rating system. The two designations for LEEDTM professional credentials include LEEDTM Accredited Professional (LEEDTM AP) or LEEDTM Green Associate (LEEDTM GA).

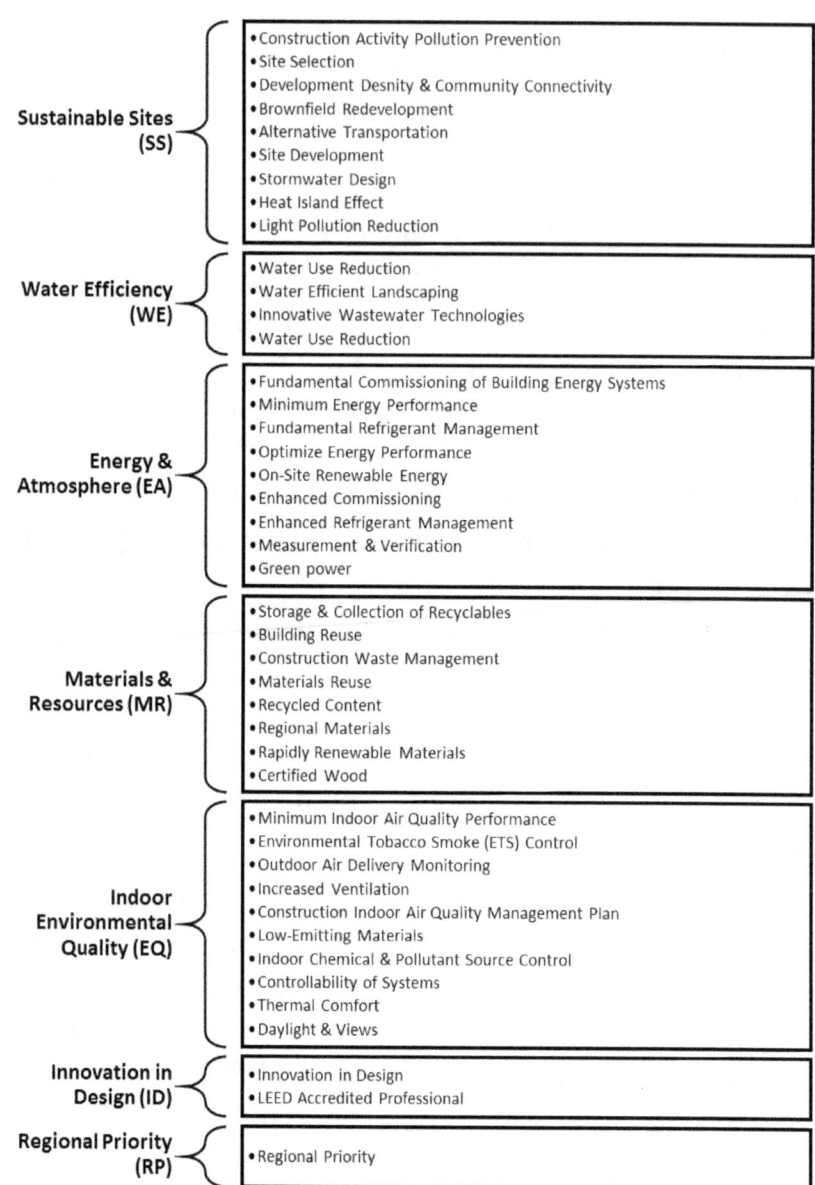

Sustainable Sites (SS)
- Construction Activity Pollution Prevention
- Site Selection
- Development Desnity & Community Connectivity
- Brownfield Redevelopment
- Alternative Transportation
- Site Development
- Stormwater Design
- Heat Island Effect
- Light Pollution Reduction

Water Efficiency (WE)
- Water Use Reduction
- Water Efficient Landscaping
- Innovative Wastewater Technologies
- Water Use Reduction

Energy & Atmosphere (EA)
- Fundamental Commissioning of Building Energy Systems
- Minimum Energy Performance
- Fundamental Refrigerant Management
- Optimize Energy Performance
- On-Site Renewable Energy
- Enhanced Commissioning
- Enhanced Refrigerant Management
- Measurement & Verification
- Green power

Materials & Resources (MR)
- Storage & Collection of Recyclables
- Building Reuse
- Construction Waste Management
- Materials Reuse
- Recycled Content
- Regional Materials
- Rapidly Renewable Materials
- Certified Wood

Indoor Environmental Quality (EQ)
- Minimum Indoor Air Quality Performance
- Environmental Tobacco Smoke (ETS) Control
- Outdoor Air Delivery Monitoring
- Increased Ventilation
- Construction Indoor Air Quality Management Plan
- Low-Emitting Materials
- Indoor Chemical & Pollutant Source Control
- Controllability of Systems
- Thermal Comfort
- Daylight & Views

Innovation in Design (ID)
- Innovation in Design
- LEED Accredited Professional

Regional Priority (RP)
- Regional Priority

Fig. 11.1. LEED™ (version 3, 2009) prerequisites and credits under each key area [*Source:* USGBC (3)]

UK's Civil Engineering Environmental Quality Assessment and Award Scheme (CEEQUAL)

The CEEQUAL was developed under the leadership of the UK's Institution of Civil Engineers (ICE) through a pan-industry, government funded project as an award-based assessment scheme for the environmental quality of civil engineering and public realm projects. It was launched for public use in June 2004. Since then, it has gained popularity in the UK construction industry and is currently managed by CEEQUAL Ltd, a company with several contributing project partners. A significantly updated version of the CEEQUAL Assessment Manual for Projects (Version 4) was launched in November 2008 which can be applied to roads, railways, airports, etc. and projects of any scale (8).

The CEEQUAL was developed with the aim of improving sustainability in civil engineering and public realm projects and to demonstrate the civil engineering industry's commitment to environmental quality and social performance. Based on a rigorous assessment of performance across 12 areas of environmental and social concern, the CEEQUAL Scheme rewards projects which demonstrate distinctive environmental and social standards beyond the legal and environmental minima (8,9).

The CEEQUAL self-assessment process involves answering a series of searching questions by a trained CEEQUAL Assessor based on the manual as well as an online assessment tool provided by CEEQUAL. Some of the typical questions include: Have measures been incorporated to conserve water and reduce water consumption during operation of the project?, Have sustainable drainage systems been incorporated in the project?, etc. The answers to these questions will help capture the scores a project deserves as well as log justifying evidence for the scores. It is recommended that the assessment process is undertaken a rolling activity through the life of the project. A CEEQUAL-appointed Verifier then reviews the assessment results and makes a recommendation to CEEQUAL for an Award (8). The different Award levels a project can achieve based on the percentage of points scored are:

✓ Pass (over 25%)

✓ Good (over 40%)

✓ Very Good (over 60%)

✓ Excellent (over 75%)

The question set contained in the CEEQUAL Assessment Manual comprise the CEEQUAL Scheme against which projects are assessed and scored. The questions are split across the 12 sections as shown in Fig. 11.2 with the associated weights.

Currently, a CEEQUAL Award is only available for projects in the UK and Ireland. An International Version of the CEEQUAL is being launched in early 2011.

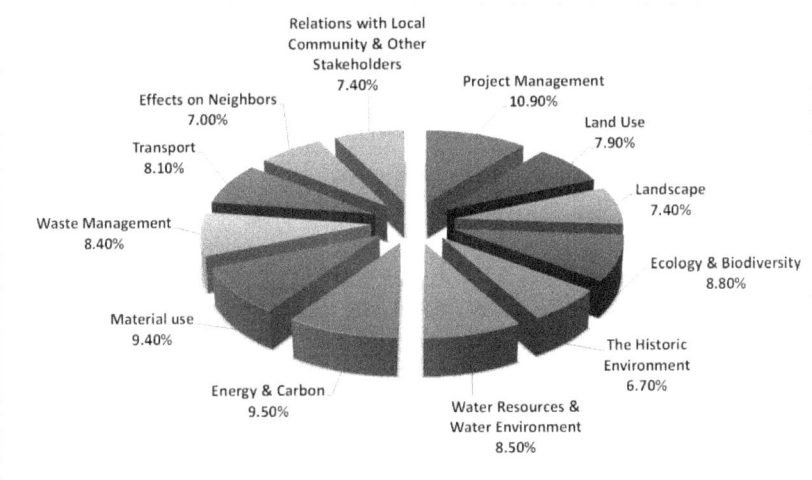

Fig. 11.2. CEEQUAL assessment question categories [*Source: CEEQUAL (8)*]

Greenroads™

The Greenroads™, jointly developed by the University of Washington (UW) and CH2M Hill, Inc., is a voluntary sustainable rating system or "performance metric" that could be used for quantifying the sustainable attributes of a roadway project (10). It defines sustainability as "a system characteristic that reflects the system's capacity to support natural laws and human values" and a Greenroad as "roadway project that has been designed and constructed to a level of sustainability that is substantially higher than current common practice." These definitions make it clear that Greenroads™ is a project-based system that best applies to the design process and construction activities of a roadway project (11). The construction activities include production of

PCC and HMA, material hauling activities, and those within the workzone.

In assessing the sustainability of a roadway project, Greenroads™ awards points for more sustainable practices/choices and can be used to define the project attributes contributing to roadway sustainability, communicate this information to stakeholders, and grant "certification" based on achieving a minimum number of points.

Similar to LEED™ system for rating buildings, Greenroads™ evaluates a roadway project using a number of criteria and awards points for each. Each sustainability practice is weighted against its overall life-time ecological, economic, and social impacts based on representative empirical data and practices with greater lifetime impact on sustainability performance get a higher point value (12).

The sustainability best practices accepted by Greenroads™ are divided into two types:

- *Mandatory "Project Requirements"* for a roadway to be considered a Green road, and

- *Voluntary best practices* which may be optionally included in a roadway project.
  - Environment & Water (stormwater, habitat, vegetation)
  - Access & Equity (modal access, culture, aesthetics, safety)
  - Construction Activities (construction equipment, quality, use)

o Materials & Resources (material extraction, processing, transport)

o Pavement Technology (pavement design, material, use, function)

Greenroads™ consists of a grand total of 118 credits in these 6 categories (plus custom credits) that can be used to achieve certification. To achieve Greenroads™ certification, a roadway project must meet the mandatory Project Requirements which includes completing an environmental review process and a LID hydrologic analysis; establishing a Quality Control (QC) plan, noise mitigation plan, waste management plan, pollution prevention plan, site maintenance plan, etc. and gain (Voluntary Credit) points for such items as habitat restoration, improving bicycle accessibility and using recycled and local materials. A blank Greenroads™ Scorecard is shown in Fig. 11.3.

The following four achievement levels (subject to revision) awarded by Greenroads™ and these levels are intended to be above and beyond what could be achieved by current common practice:

✓ *Certified:* All Project Requirements + 32-42 Voluntary Credit points (30-40% of total)

✓ *Silver:* All Project Requirements + 43-53 Voluntary Credit points (40-50% of total)

✓ *Gold:* All Project Requirements + 54-63 Voluntary Credit points (50-60% of total)

✓ *Evergreen:* All Project Requirements + 64+ Voluntary Credit points (>60% of total)

Project Name
Location (City, State)
Project Manager
Project Budget _____ million (USD)
Current Status (Circle) Planning   Design   Construction   Complete

Project Requirements (PR)	PR Max: 11		
No. Title	Y	?	N
PR-1 Environmental Review Process	Req		
PR-2 Lifecycle Cost Analysis	Req		
PR-3 Lifecycle Inventory	Req		
PR-4 Quality Control Plan	Req		
PR-5 Noise Mitigation Plan	Req		
PR-6 Waste Management Plan	Req		
PR-7 Pollution Prevention Plan	Req		
PR-8 Low-Impact Development	Req		
PR-9 Pavement Management System	Req		
PR-1 Site Maintenance Plan	Req		
PR-11 Educational Outreach	Req		

Environment & Water (EW)	EW Max: 21		
	Y	?	N
EW-1 Environmental Management System	2		
EW-2 Runoff Flow Control	1 - 3		
EW-3 Runoff Quality	1 - 3		
EW-4 Stormwater Cost Analysis	1		
EW-5 Site Vegetation	1 - 3		
EW-6 Habitat Restoration	3		
EW-7 Ecological Connectivity	1 - 3		
EW-8 Light Pollution	3		

Access & Equity (AE)	AE Max: 30		
	Y	?	N
AE-1 Safety Audit	1 - 2		
AE-2 Intelligent Transportation Systems	2 - 5		
AE-3 Context Sensitive Solutions	5		
AE-4 Traffic Emissions Reduction	5		
AE-5 Pedestrian Access	1 - 2		
AE-6 Bicycle Access	1 - 2		
AE-7 Transit & HOV Access	1 - 5		
AE-8 Scenic Views	1 - 2		
AE-9 Cultural Outreach	1 - 2		

Construction Activities (CA)	CA Max: 14		
No. Title	Y	?	N
CA-1 Quality Management System	2		
CA-2 Environmental Training	1		
CA-3 Site Recycling Plan	1		
CA-4 Fossil Fuel Reduction	1 - 2		
CA-5 Equipment Emission Reduction	1 - 2		
CA-6 Paving Emission Reduction	1		
CA-7 Water Use Tracking	2		
CA-8 Contractor Warranty	3		

Materials & Resources (MR)	MR Max: 23		
	Y	?	N
MR-1 Lifecycle Assessment	2		
MR-2 Pavement Reuse	1 - 5		
MR-3 Earthwork Balance	1		
MR-4 Recycled Materials	1 - 5		
MR-5 Regional Materials	1 - 5		
MR-6 Energy Efficiency	1 - 5		

Pavement Technologies (PT)	PT Max: 20		
	Y	?	N
PT-1 Long-Life Pavement	5		
PT-2 Permeable Pavement	3		
PT-3 Warm Mix Asphalt	3		
PT-4 Cool Pavement	5		
PT-5 Quiet Pavement	2 - 3		
PT-6 Pavement Performance Tracking	1		

Custom Credit (CC)	CC Max: 10		
	Y	?	N
CC-1 _____	1 - 5		
CC-2 _____	1 - 5		
CC-3 _____	1 - 5		

All 11 PR Met?			
Greenroads Total (Max 118)			

Fig. 11.3. Blank Greenroads™ scorecard (subject to revision) [*Source:* Greenroads™ (10)]

GreenLITES

The GreenLITES (Green Leadership in Transportation Environmental Sustainability) certification program was developed by the New York State Department of Transportation (NYSDOT) to help integrate sustainability practices into transportation (13).

Initially designed to assess environmental sustainability on individual capital projects, it later evolved into a holistic approach incorporating the "triple bottom line" of sustainability and is now used for the planning and assessment of transportation projects, operational activities, maintenance programs, and regional investments. It is primarily being used as internal management program and a self-assessment tool for NYSDOT to measure its sustainability performance, recognize and encourage sustainability practices, and identify areas of improvement (14).

GreenLITES is modeled after the LEED™ sustainability rating system for green building practices as well as the Greenroads™ program. It mainly emphasizes the design aspect of transportation projects and evaluates NYSDOT project designs for over 178 sustainable practices in the following five categories (see Fig. 11.4 for a screenshot of blank GreenLITES scorecard):

✓ *Sustainable Sites:* Alignment selection, CSS, land use community planning, protect, restore, enhance wildlife habitat and plant communities

✓ *Water Quality:* Stormwater management and runoff pollutant management

✓ *Materials and Resources:* Reuse of materials, recycled content, locally provided material, bioengineering techniques, and hazardous materials minimization

✓ *Energy and Atmosphere:* Improve traffic flow, reduce electrical consumption, reduce petroleum consumption, improve pedestrian/bike access, noise abatement, and reduce stray light

✓ *Innovation/Unlisted:* Designs that significantly build upon GreenLITES categories or incorporate significant innovations in transportation environmental sustainability

Similar to Greenroads™ and LEED™ programs, GreenLITES provides a certification level based on points obtained on a project with the highest certification level (Evergreen) awarded to designs which clearly advance the state of sustainable transportation solutions (15):

✓ *Non-certified* (0 – 14 points)

✓ *Certified* (15 – 29 points)

✓ *Silver* (30 – 44 points)

✓ *Gold* (45 – 59 points)

✓ *Evergreen* (60+ points)

GreenLITES Project Environmental Sustainability Rating System Scorecard					
				POINTS	
CATEGORY	ID	DESCRIPTION		Available	Scored
(BMPs)	W-2d	Minimizing the increase in impervious area due to the project.		1	
	W-2e	Including grass channels, where appropriate.		1	
M-1 Reuse of Materials	M-1a	Specify that 75% or more of topsoil removed for grading is reused on site.		2	
	M-1b	Cuts and fills balanced to within 10 percent.		2	
	M-1c	Reuse spoil within project corridor to minimize material in and out of site.		2	
	M-1d	Specify rubblizing or crack and seating of Portland Cement Concrete pavement.		2	
	M-1e	Reuse of previous pavement as subbase for full-depth reconstruction projects.		2	
	M-1f	Arranging for the reuse of excess excavated material, asphalt pavement millings, or demolished concrete by another municipality or state agency.		2	
	M-1g	Specify the processing of demolished concrete to reclaim scrap metals and to create a usable aggregate material.		2	
	M-1h	Salvaging removed trees for lumber or uses other than standard wood-chipping.		2	
	M-1i	Use surplus excavated material on nearby state highways for slope flattening to eliminate guide rail or as fill in areas designated by Park officials as acceptable for spoil disposal.		2	
	M-1j	Use surplus excavated material, demolished concrete, or millings at nearby abandoned quarries to help fulfill an approved DEC reclamation plan.		2	
	M-1k	Specify that 50% or more of topsoil removed for grading is reused on site.		1	
	M-1l	Design the project so that "cut-and-fills" are balanced to within 25 percent.		1	
	M-1m	Reuse (i.e. remove and reset versus remove and replace) of granite curbing.		1	
	M-1n	Reuse of stone veneer or other elements of the previous structure.		1	
	M-1o	Design on-site location for chipped wood waste disposal from clearing & grubbing.		1	
	M-1p	Specifying the recycling of chipped untreated wood waste for use as mulch and/or ground cover.		1	
	M-1q	Project documents make scrap metals available for reuse or recycling.		1	
	M-1r	Identify approved, environmentally acceptable and permitted sites in the contract documents for the disposal of surplus excavated material.		1	
		Obtain and implement a project specific DEC Beneficial Use Determination for			

(Materials & Resources (M) — vertical label on left side of table)

Fig. 11.4. Screenshot of GreenLITES project environmental sustainability rating system scorecard [*Source:* NYSDOT (15)]

The 'Route 347 Safety and Mobility Project (Long Island, New York)' is cited as an example with GreenLITES 'Evergreen' certification (15). The 'green' elements proposed in the vision plan included: solar-LED bus stop, bio-swales, solar lighting, energy-efficient traffic signals, rain garden, traffic calming measures, green 'living' walls as alternatives to traditional tall concrete sound walls, etc (see Fig. 11.5).

Fig. 11.5. Greenway concept for Route 347 Safety and Mobility Project in Long Island, New York [*Source:* NYSDOT (16)]

Ontario Ministry of Transportation's (MTO) GreenPave

GreenPave is a points-based rating system developed by Ontario's Ministry of Transportation (MTO) for encouraging environment-friendly elements of road paving design. Although, GreenPave is modeled after Greenroads™ and GreenLITES rating systems, it is customized for Ontario's conditions and focuses strictly on pavement, not the entire right-of-way. Also, GreenPave is exclusively used by Ontario's MTO to rate environmental sustainability at project level for internal management similar to the use of GreenLITES by NYSDOT.

Based on a joint effort by the University of Waterloo, Centre of Pavement and Transportation Technology (UW CPATT) and MTO, the Green Pave project began in September 2008 and concluded in April 2010 with the goal of developing a sustainable pavement framework for pavement engineering practice in Ontario. This was carried out under a research project entitled "Quantifying Pavement Sustainability" sponsored by the MTO's Highway Infrastructure Innovation Funding Program (HIIFP) (17).

Every project is assessed using four categories in Green Pave and a project can gain up to a total of 36 credits divided under three themes: design credits, construction credits, and innovation credits. The four GreenPave points categories are:

✓ *Pavement Technologies:* Long-life pavement designs, permeable pavements, quiet pavements, and cool pavements.

✓ *Materials and Resources:* Recycled content, reuse of pavement, local materials, and construction quality.

✓ *Energy and Atmosphere:* Reduce energy consumption, GHG emission reduction, improve rolling resistance, pollution reduction.

✓ *Innovation and Design Process:* Pollution reduction, innovation in design, and exemplary process.

While design credits are assessed during pavement design assessment, construction credits are awarded post construction and innovation credits are awarded for sustainable practices not identified in GreenPave. A sample GreenPave scorecard using MS-Excel® is shown in Fig. 11.6 summarizing the evaluation results of a sample project. Depending on the total number of points earned, four certification levels are proposed in GreenPave (18):

✓ *Bronze* (7-10 points)

✓ *Silver* (11-14 points)

✓ *Gold* (15-19 points)

✓ *Trillium* (20+ points)

In the development of GreenPave program, typical environmental and economic savings between different pavement technologies and of different project alternatives were quantified and examined from both project-level and life cycle perspectices. Environmental quantification of energy and GHG emissions completed using the PaLATE LCA software tool (19) showed that the in-place recycling processes were the most environmentally friendly (17).

The Engineering Development Program (EDP) interns and staff in the Pavements and Foundations Section of the Materials Engineering and Research Office at MTO developed the GreenPave rating system and beta tested it. It is currently being fine-tuned in consultation with regional Geotechnical sections and industry partners such as the Ontario Hot Mix Producers Association (OHMPA) and the

Ready Mixed Concrete Association of Ontario (RMCAO) (18).

GreenPave Project Checklist

WP: 476-98-00
Alternative Description: IPP to 160mm, 90mm HM
LCCA of Alternative 1: $96.003 per 2-lane Km
$2,346,894

MAX POINTS	GreenPave CREDIT NUMBER	GreenPave CREDIT NAME	TOTAL GreenPave PROJECT POINTS
9	**Pavement Technologies**		**1 Points**
	Credit PT - 1	Long-Life Pavement	0
	Credit PT - 2	Permeable Pavements	0
	Credit PT - 3	Noise Mitigation	1
	Credit PT - 4	Cool Pavements	0
14	**Materials & Resources**		**8 Points**
	Credit MR - 1.1	Recycled Content - Part 1	
	Credit MR - 1.2	Recycled Content - Part 2	
	Credit MR - 1.3	Recycled Content - Part 3	
	Credit MR - 1.4	Recycled Content - Part 4	4
	Credit MR - 2	Reuse of Pavement	2
	Credit MR - 3	Local Materials	2
	Credit MR - 4	Construction Quaility	0
9	**Energy & Atmosphere**		**5 Points**
	Credit EA - 1	Reduce Energy Consumption	3
	Credit EA - 2	GHG Emission Reduction	2
	Credit EA - 3	Rolling Resistance	0
	Credit EA - 4	Pollution Reduction	0
4	**Innovation & Design Process**		**0 Points**
	Credit I - 1	Innovation in Design	0
	Credit I - 2	Exemplary Process	0
36	**Total Project Points**		**14 Points**

Certified 7-10 points   Silver 11-14 points   Gold 15-19 points   Trillium 20-36 points

Your design is GreenPave SILVER!

Summary | Alternative 1 | Alternative 2 | Alternative 3 | MR - 1.4 Calcs

Fig. 11.6. A sample GreenPave scorecard (beta version) in Ms-Excel® [*Source*: (17)]

FHWA's INVEST

The FHWA's Infrastructure Voluntary Evaluation Sustainability Tool (INVEST) "identifies characteristics of sustainable highways and provides procedures and techniques to help agencies and organizations apply and integrate sustainability best practices into highway and other roadway projects and programs within system planning, project development, and operations and maintenance"(2). By establishing standard and qualitative sustainability measures through this program, FHWA's goal is to encourage increased application of sustainability principles in highway projects and establish an implementable baseline foundation for highway sustainability. All roadway projects are considered and not just highways.

According to FHWA, "a sustainable highway should satisfy life cycle functional requirements of societal development and economic growth while reducing negative impacts to the environment and consumption of natural resources." Apart from the "triple bottom line" (ecology, economy, and equity) principles of sustainability, three other essential components were considered in developing the tool: context, performance, and education. Context can be incorporated by considering the project's scope (i.e., size and scale) and system boundaries (i.e, life cycle). The performance component can be incorporated by considering the expectations from the project such as design life, metrics of performance, assessment of risk, and unintended consequences. Finally, the education component is required to propagate sustainable practices through educational outreach programs and drive progress and implementation of sustainable solutions (2).

At the outset, the FHWA's INVEST appears similar to the Greenroads™ program. In fact, the Greenroads™ team was assigned the task of developing the self-evaluation tool on behalf of FHWA under contract. However, the FHWA's tool is a separate and distinct effort from the Greenroads™ program. The beta version of the FHWA's Sustainable Highways Self-Evaluation Tool released in Fall 2010 uses 68 credits broken down into following three categories to help agencies assess their sustainability performance:

✓ System Planning (SP) – see Fig. 11.8

✓ Operations and Maintenance (OM) – see Fig. 11.9

✓ Project Development (PD) – see Fig. 11.10

The Sustainable Highways Self-Evaluation Tool is different from the other rating systems discussed in this chapter in that each of these three categories has its own credits that can be used to independently evaluate a given project. For instance, in evaluating the sustainability of a highway project, the PD category is used and not the other two categories. Eighteen defined sustainability benefits (those shown in Fig. 11.7 plus optimization of habitat and land use) were used in developing the tool and it is possible to trace back each of the benefits to one or more of the sustainability principles.

REDUCE	IMPROVE	INCREASE	CREATE
•Raw materials use •Fossil energy use •Water use •Air emissions •Wastewater •Sediment and solid waste	•Safety and human health •Access and mobility •Transportation choices •Economic prosperity •Business practice	•Lifecycle savings •Lifecycle service •Awareness •Aesthetics	•New information •Energy

Fig. 11.7. Benefits of Sustainable Highways [*Source:* FHWA (2)]

Credit	Title	Score
SP-1	Comprehensive and Integrated Planning	0
SP-2	Environmental Management System	0
SP-3	Context Sensitive Solutions (CSS)	0
SP-4	Equity Analysis	0
SP-5	Land Use Planning Integration	0
SP-6	Multimodal Transportation	0
SP-7	Professional Development	0
SP-8	Travel Demand Management	0
SP-9	Safety Planning	0
SP-10	Air Quality	0
SP-11	Greenhouse Gas Emissions	0
SP-12	Climate Change Effects	0
SP-13	Noise Reduction Management Plan	0
SP-14	Financial Sustainability	0
	System Planning and Processes Score	**0**

Fig. 11.8. FHWA Sustainable Highways Self-Evaluation Scorecard: System Planning and Processes [*Source:* FHWA (20)]

Credit	Title	Score
OM-1	Pollution Prevention Plan	0
OM-2	Pavement Management System	0
OM-3	Bridge Management System	0
OM-4	Paved Surfaces Management System	0
OM-5	Traffic Control Infrastructure Maintenance	0
OM-6	Cleaning and Litter	0
OM-7	Roadside Infrastructure Maintenance	0
OM-8	Snow and Ice Control	0
OM-9	Mobility	0
OM-10	Safety	0
OM-11	Renewable Energy Use	0
OM-12	Sustainability Plan	0
OM-13	Alternative Fuel Fleet	0
OM-14	Recycle and Re-use	0
OM-15	Ecological Connectivity	0
	TSMOM Score	0

Fig. 11.9. FHWA Sustainable Highways Self-Evaluation Scorecard: Transportation Systems, Management, Operations and Maintenance Scorecard [*Source:* FHWA (20)]

The FHWA's INVEST tool is undergoing pilot testing in the summer and Fall of 2011 and Version 1.0 is expected to be released later in 2012 after incorporating the feedback received during the pilot testing.

Sustainable Sites Initiative™ (SITES™)

The Sustainable Sites Initiative™ (SITES™) is "an interdisciplinary effort by the American Society of Landscape Architects, the Lady Bird Johnson Wildflower Center at The University of Texas at Austin and the United States Botanic Garden to create voluntary national guidelines and performance benchmarks for sustainable land design, construction and maintenance practices." The projects that can be assessed by SITES™ range from corporate campuses and transportation corridors to public parks and single-family residences (21).

The Sustainable Sites Initiative: Guidelines and Performance Benchmarks 2009 report (22), a result of more than four years of work by a diverse group of experts in areas ranging from soils to human health, describes the criteria for sustainable land practices and the rating system used for measuring and rewarding projects that protect, restore, and regenerate ecosystem services. A companion document, The Case for Sustainable Landscapes, describes the purpose and principles in SITES™, arguments supporting the adoption of sustainable land practices, and the science behind the SITES™ performance benchmarks.

The SITES™ Guidelines (22) is modeled after the LEED™ rating system and is built on the central principle that any landscape has the potential to improve as well as to regenerate the natural benefits and services provided by ecosystems in their undeveloped state. Ecosystem processes are defined as the physical, chemical, and biological actions or events that link living organisms (vegetation, soil organisms, etc.) and their environment (bedrock, water, and air). Ecosystems services are benefits derived by humankind from ecosystem processes and such benefits can include products like clean drinking water and processes like decomposition of wastes (see Fig. 11.11). The SITES Guidelines identifies a list of ecosystem services that a sustainable site can strive to protect or regenerate by adopting sustainable land development and management practices.

Credit	Title	Score
PD-1	Cost Benefit Analysis	0
PD-2	Highway and Traffic Safety	0
PD-3	Context Sensitive Solutions (CSS)	0
PD-4	Lifecycle Assessment	0
PD-5	Lifecycle Cost Analysis	0
PD-6	Freight Mobility	0
PD-7	Educational Outreach	0
PD-8	Habitat Restoration	0
PD-9	Runoff Flow Control	0
PD-10	Runoff Quality	0
PD-11	Ecological Connectivity	0
PD-12	Low Impact Development	0
PD-13	Recycled Materials	0
PD-14	Renewable Energy	0
PD-15	Site Vegetation	0
PD-16	Pedestrian Access	0
PD-17	Bicycle Access	0
PD-18	Transit & HOV Access	0
PD-19	Historical, Archaeological, and Cultural Preservation	0
PD-20	Scenic, Natural, or Recreational Qualities	0
PD-21	Low-Emitting Materials	0
PD-22	Energy Efficiency	0
PD-23	Traffic Systems, Management and Operations (TSMO)	0
PD-24	Long-Life Pavement	0
PD-25	Pavement and Structure Reuse	0
PD-26	Stormwater Cost Analysis	0
PD-27	Thermal Pavement	0
PD-28	Contractor Warranty	0
PD-29	Stormwater Pollution Prevention Plan	0
PD-30	Environmental Training	0
PD-31	Equipment Emission Reduction	0
PD-32	Fossil Fuel Reduction	0
PD-33	Construction Noise Mitigation	0
PD-34	Quality Control Plan	0
PD-35	Reduced Energy Materials	0
PD-36	Waste Management	0
PD-37	Earthwork Balance	0
PD-38	Environmental Management System	0
PD-39	Tracking Environmental Commitments	0
	Project Development Score	**0**

Fig. 11.10. FHWA Sustainable Highways Self-Evaluation Scorecard: Project Development [*Source:* FHWA (20)]

The SITES™ rating system assesses the performance of a specific site on a 250-point scale with points awarded based on credits for site selection, the use of materials, restoration of soils and vegetation, and sustainable practices in construction and maintenance. Projects that have achieved all the prerequisites and at least 40% of total points are recognized as achieving "One Star" certification. A project may achieve up to "Four Stars" by implementing sustainability practices and completing additional credits. The program is currently under the pilot project phase and future refinements to the rating system are anticipated after its completion (22). The following are the nine sections under which the (pilot) project perquisites and credits are organized:

✓ Site Selection

✓ Pre-Design Assessment and Planning

✓ Site Design—Water

✓ Site Design—Soil and Vegetation

✓ Site Design—Materials Selection

✓ Site Design—Human Health and Well-Being

✓ Construction

✓ Operations and Maintenance

✓ Monitoring and Innovation

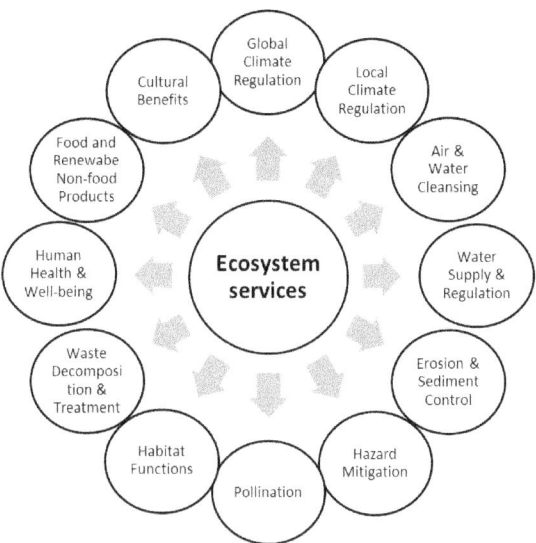

Fig. 11.11. List of ecosystem services a sustainable site can strive to protect or regenerate [*Source:* Sustainable Sites Initiative™ (22)]

BE²ST-in-Highways™

The Wisconsin DOT and the University of Wisconsin-Madison Civil Engineering Department developed the BE²ST in-Highways™ system to provide a quantitative methodology for rating the sustainable highway construction benefits (23). Using this system, alternative highway construction strategies can be evaluated and compared in terms of energy and GHG emissions over the entire life cycle using quantitative assessment techniques (24). A unique feature of this system, according to the developers, is that it allows for a transparent linkage between the project rating and the sustainable practices employed in design and construction, thus reducing the potential for 'gaming' of the rating system, a common problem encountered in building sustainability rating systems (25).

The BE²ST in-Highways™ system uses two layers of indicators suggested by a previous research study: mandatory screening and judgment indicators. The mandatory screening layer is similar to the mandatory Project Requirements in the Greenroads™ system. It uses regulatory indicators (related to local regulations, public demands, mandatory requirements to meet project needs, etc.) and project specific indicators (related to cultural and aesthetic concerns) first to eliminate from further assessment those projects which do not meet the given criteria requirements. The judgment layer uses the following economic and environmental sustainability indicators/criteria with established target values (26):

✓ Greenhouse Gas Emissions

✓ Energy Use

✓ Waste Reduction (Including *ex situ* materials)

✓ Waste Reduction (Recycling *in situ* materials)

✓ Water Consumption

✓ Social Carbon Cost (SCC) Saving

✓ Life Cycle Cost

✓ Traffic Noise

✓ Hazardous Waste

In an actual highway project evaluation, weights (since there are two targets for each criterion) and credits are selected for each of the criterion by stakeholders using the Analytical Hierarchy Process (AHP) (27). A separate software package is provided for computing the weights using AHP. The sustainability of the highway project is judged in terms of quantitative difference between a reference design (a conventional highway construction design) and the proposed alternative design(s) resulting in relative score(s). An equally weighted system will consist of 2 points for each of the 9 criteria resulting in a total of 18 points which is the default BE²ST in-Highways™ system. The total score for an alternative design is calculated as the sum of points obtained as a fraction of 18 total points (percentage). Three certification levels are used depending on the percentage:

- ✓ Bronze (50%)
- ✓ Silver (75%)
- ✓ Gold (90%)

Recent pilot project evaluation using the BE²ST in-Highways™ system indicated that the greatest environmental benefit (in terms of reduced CO_2 emissions) and energy savings result from reduction of raw materials used in highway construction or replacement of conventional construction materials with recycled materials (23).

Alberta's Green Guide for Roads

The Green Guide for Roads was originally developed as an internal concept document by Stantec Consulting Ltd. for setting a benchmark in sustainable transportation practice as well as to use it as a marketing tool for potential clients (28). Subsequent revisions have been made to it by comparing it with other rating systems such as Greenroads™ and GreenLITES and the efforts have resulted in a functional, working document (29).

The Alberta Green Guide for Roads is structured after the LEED™ rating system and is broken up into seven categories (and subcategories) with each category containing prerequisites which must be met for certification as well as optional credits (see Fig. 11.12).

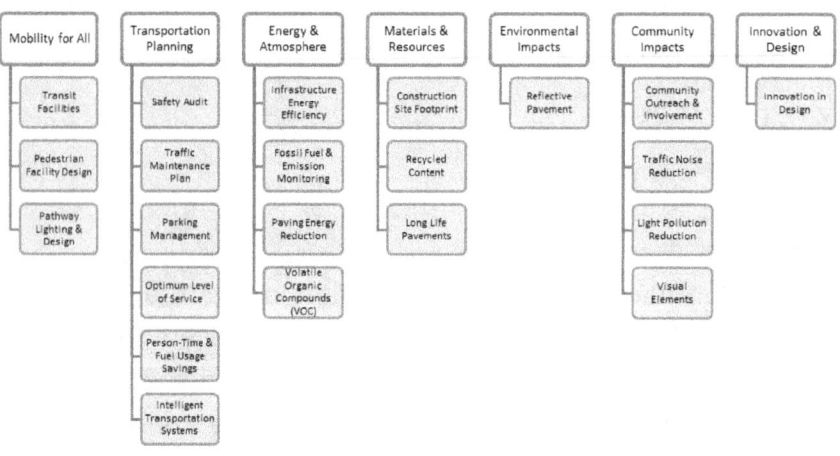

Fig. 11.12. Alberta's Green Guide for Roads Categories
[*Source:* (28)]

The *Mobility for All* category promotes alternative transportation choices by making roads more bicycle and pedestrian friendly (similar to the Complete Streets policy) and encourages the use of public transit. *Energy & Atmosphere* encourages energy reduction and conservation during construction related activities and promotes alternative pavement designs that reduce emissions and improve safety of work environment. *Materials & Resources* limits the construction site footprint by preventing the use of virgin materials and encouraging recycling and reuse, the use of local materials, and long-life pavement designs. *Community Impacts* increases community involvement in the planning stages of the project and enhances visual elements while minimizing traffic noise and light pollution (28).

Transportation Association of Canada's Green Guide for Roads Task Force

The Transportation Association of Canada's (TAC) Urban Transportation Council established a Task Force "to develop a set of self-evaluation performance measurement criteria incorporating sustainable/green principles and environmental stewardship that may be applied to all aspects of the roadway over its lifecycle." This will include providing guidance for green roads in both urban and rural settings starting from roadway planning to design to construction to maintenance and operation and other life cycle assessment activities. All major categories of roadways will be considered including freeway/expressway, arterial, collector, and local roads with the goal of promoting sustainable growth, safe, long-lasting roadway infrastructure and alternative multi-modal solutions within corridors (30).

The TAC Task Force's goal is to establish a Green Guide for the Canadian transportation industry by expanding and adapting Stantec's Green Guide for Roads. Based upon the work completed by the Task Force, a TAC sponsored project was initiated in May 2010 with the goal of developing a Green Guide for Roads in Canada and the project is expected to be complete in September 2011 (31). The final draft of the guide is expected to be ready by May 2012 for publication. On September 26th, 2010, the TAC Task Force officially transferred the Guide development to the Project Steering Committee and requested for a motion to formally disband the Task Force (32).

A rating system similar to the LEED™ system has been used as a starting point for the evolution of the Green Guide for Roads which has also adopted various criteria and ideas mentioned in Greenroads™ and GreenLITES rating system manuals. A Working Template of the Green Guide for Roads is currently available for download from the TAC's project website which lists the following 13 application areas (benchmarks) with goals and sub-goals defined (33):

✓ Community Interface

✓ Environmental Footprint

✓ Mobility Choices

✓ Intersections and Driveways

✓ Hard Surfaces

✓ Landscaping

✓ Street Furnishings

✓ Drainage

✓ Safety

✓ Energy Consumption

✓ Construction

✓ Operation and Maintenance

✓ Services and Utilities

Institute for Sustainable Infrastructure (ISI) Rating System

The American Society of Civil Engineers (ASCE) has been working jointly with the American Council of Engineering Companies (ACEC) and the American Public Works Association (APWA) in developing a sustainable infrastructure rating structure similar to the LEED™ rating system for buildings. As a result, the Institute for Sustainable Infrastructure (ISI) was established (similar to USGBC for LEED™) as a joint-venture, non-profit with ASCE, ACEC and APWA as founding members and with the mission of providing "sustainability products and services that will transform infrastructure design, construction, and operation by taking into account the Triple Bottom Line" (34).

The ISI infrastructure rating system is expected to provide a uniform means of assessing sustainability benefits for practitioners, owners and regulators. It is considered as the first ever "comprehensive" sustainable infrastructure rating system developed in North America. While LEED™ applies to only building projects, at least as of now, the ISI rating tool is intended to assess the extent to which infrastructure systems like highways, bridges, wastewater treatment facilities, and power grids follow the principles of sustainable development (35). Currently, there are no U.S.

programs similar to CEEQUAL (8), the UK based civil infrastructure rating system.

The ISI rating system will provide performance-based outcomes rather than prescriptive with a two-pronged purpose: one is to ensure that the project makes a pathway contribution by doing the right thing with a community-focused approach, and the other is to ensure that the project makes performance contribution by do things right. It will be scalable for accommodating projects of different sizes and degrees of complexity as well as adaptable for specific needs and circumstances (36).

The ISI partners will be offering sustainability training programs and related educational materials as well as develop a certification program to further promote sustainable infrastructure design and practice. The scope of the ISI rating system includes planning, design, construction and operation and a range of scores are possible for sustainable performance – from conventional to sustainable to restorative (37).

The ISI infrastructure rating system is currently under full-scale piloting and is expected to be released for public use in the late Spring 2011 as a voluntary web-based product. The Version 1.0 of the ISI rating system will have 10 sections (see Fig. 11.13) with targeted assessment questions for each section arranged in sub-sections. The awards process will begin in October 2011 (36).

I-LAST™

I-LAST™ (The Illinois-Livable and Sustainable Transportation Rating System and Guide) is the result of two years of cooperative effort between the Illinois DOT, Illinois Road and Transportation Builders Association (IRTBA), and the American Council of Engineering Companies of Illinois (ACEC-IL). The I-LAST™ Manual (Version 1.0) released on January 20, 2010 is considered a living document which will be revised and updated as the state-of-the-art evolves with inputs from industry users. Similar to the Greenroads™ and BE²ST in-Highways™ systems, the purpose of the I-LAST™ guide is to recognize and encourage the use of sustainable practices in the transportation industry as well as to establish a simple performance metric system that can be used in evaluating transportation projects with respect to livability, sustainability, and effect on the natural environment (38).

No.	Section	Weight (%)
1	Pathway	12.6
2	Project Strategy & Management	10.6
3	Community: Long & Short Term Effects	10.7
4	Land Use & Restoration	8.9
5	Landscapes	7.0
6	Ecology & Biodiversity	8.8
7	Water Resources & Environment	11.5
8	Energy & Carbon	11.7
9	Resource Management Including Waste	8.2
10	Transportation	10.0
	TOTAL	100%

Fig. 11.13. ISI rating system primary criteria (beta version) [Source: ACEC (36)]

The I-LAST™ guide includes a variety of potential sustainable practices and features against which a highway project can be evaluated using I-LAST™'s scorecard. The I-LAST™'s sustainable practices are divided into eight categories as shown in Fig. 11.14.

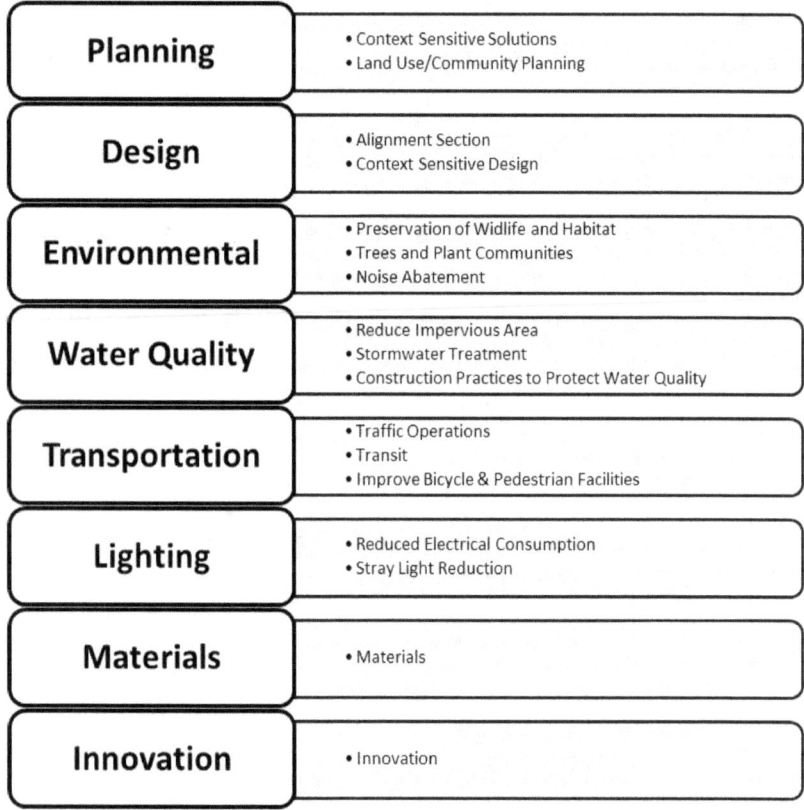

Fig. 11.14. I-LAST™ categories for sustainable practices [*Source:* IDOT et al. (38)]

Other Green Highway Rating Systems

✓ *SD-Flex* (39): A general framework for the design and construction of sustainable flexible pavement system by employing a point system to award a rating that quantifies the sustainability of the structure.

✓ *Sustainable Corridor Rating System (SCRS)* (40): A green rating system for urban transportation corridors.

✓ *Green Highway Construction Rating System in Mississippi* (41): A green highway construction rating system for Mississippi greenway based on the Greenroads™ and GreenLITES initiatives.

Useful Online Resources

USGBC LEED™

http://www.usgbc.org/LEED/

UK CEEQUAL

http://www.ceequal.com/

Greenroads™

http://www.greenroads.us/

New York State DOT GreenLITES

https://www.nysdot.gov/programs/greenlites

Ontario Ministry of Transportation (MTO) GreenPave

http://www.mto.gov.on.ca/english/transtek/roadtalk/rt16-1/#a6

FHWA's Sustainable Highways Self-Evaluation Tool

http://www.sustainablehighways.org/

Sustainable Sites Initiative™ (SITES™)

http://www.sustainablesites.org/

Wisconsin BE²ST-in-Highways™

http://www.rmrc.unh.edu/Outreach/docs/Lee,Edil,Benson,Tinjum.pdf

Alberta's Green Guide for Roads

http://greenalberta.ca/roads/

Transportation Association of Canada Green Guide for Roads

http://www.tac-atc.ca/english/projects/greenguide.cfm

Institute for Sustainable Infrastructure (ISI) Rating System

http://www.asce.org/Sustainability/ISI-Rating-System/

Illinois I-LAST™

http://www.dot.state.il.us/green/projects.html

References

1. Bryce JM. Exploring Green Highways. ASTM Standardization News [Internet]. 2008 [cited 2011 Apr 3];36(5):38-43. Available from: http://www.wise-intern.org/journal/2008/JamesBryceFinal.pdf

2. Federal Highway Administration (FHWA). FHWA Sustainable Highways Self-Evaluation Tool [Internet]. 2011 [cited 2011 Feb 6];Available from: https://www.sustainablehighways.org/1/home.html

3. USGBC. LEED 2009 For New Construction and Major Renovations [Internet]. Washington D.C.: U.S. Green Building Council; 2009 [cited 2011 Feb 15]. Available from: http://www.usgbc.org/ShowFile.aspx?DocumentID=5546

4. Zimmerman G. LEED: A Scorecard for Green Buildings [Internet]. 2004 [cited 2011 Feb 16]. Available from: http://www.facilitiesnet.com/green/article/LEED-A-Scorecard-for-Green-Buildings--2020

5. USGBC. LEED Version 3 [Internet]. [cited 2011 Feb 16];Available from: http://www.usgbc.org/DisplayPage.aspx?CMSPageID=1970

6. Kats GH. Green Building Costs andFinancial Benefits [Internet]. Westborough, MA: Massachusetts

Technology Collaborative; 2003 [cited 2011 Feb 15]. Available from: http://www.nhphps.org/docs/documents/GreenBuildin gspaper.pdf

7. USGBC. What is LEED? [Internet]. Urban Green Council. 2008 [cited 2011 Mar 28];Available from: http://www.urbangreencouncil.org/leed/leed-standard.html

8. CIRIA & Crane Environmental. CEEQUAL: Background [Internet]. CEEQUAL: The Assessment and Awards Scheme for Improving Sustainability in Civil Engineering and the Public Realm. [cited 2011 Mar 28];Available from: http://www.ceequal.co.uk/background.htm

9. Thompson P. Teaching sustainability in civil engineering using Ceequal. Proceedings of the ICE - Engineering Sustainability [Internet]. 2010 [cited 2011 Mar 28];163(4):209-217. Available from: http://www.icevirtuallibrary.com/content/article/10.168 0/ensu.2010.163.4.209

10. University of Washington. The Greenroads Rating System [Internet]. [cited 2011 Mar 28];Available from: http://www.greenroads.us/

11. Muench ST, Anderson JL, Hatfield JP, Koester JR, Soderlund M. Greenroads Rating System v1.5. Seattle, WA: University of Washington;

12. Söderlund M, Muench ST, Willoughby K, Uhlmeyer J, Weston J. Green roads: A sustainability rating system for roadways [Internet]. In: Transportation Research Board 2008 Annual Meeting. Washington, D.C.:

Transportation Research Board (TRB); 2008 [cited 2011 Mar 28]. p. 21.Available from: http://itre.ncsu.edu/ADC10/PDFs/2008_Winter_Confere nce/Presentations/08-0803.pdf

13. NYSDOT. GreenLITES [Internet]. New York State Department of Transportation (NYSDOT). [cited 2011 Mar 28];Available from: https://www.nysdot.gov/programs/greenlites

14. McVoy GR, Nelson DA, Krekeler P, Kolb E, Gritsavage JS. Moving Towards Sustainability:New York State Department of Transportation's GreenLITES Story [Internet]. In: Proceedings of the Green Streets and Highways 2010 Conference. Denver, Colorado: ASCE Press; 2010 [cited 2011 Feb 28]. p. 20.Available from: https://www.nysdot.gov/programs/greenlites/repositor y/ASCE%20GreenLITES%20Final%20Paper%207-12- 10.pdf

15. NYSDOT. GreenLITES Project Design Certification Program [Internet]. 2008 Sep [cited 2011 Mar 28];Available from: https://www.nysdot.gov/programs/greenlites/repositor y/Green%20LITES%20Certification%20Program%20- %20Full%20Doc%20-%20Final.pdf

16. NYSDOT. Vision Plan for a Green Route 347 - Balancing The Transportation System [Internet]. New York State Department of Transportation (NYSDOT). 2009 [cited 2011 Mar 28];Available from: https://www.nysdot.gov/regional- offices/region10/projects/ny-347/news

17. Chan PCP. Quantifying Pavement Sustainability For Ontario Highways [Internet] [M.S. Thesis]. [Ontario,

Canada]: University of Waterloo; 2010 [cited 2011 Mar 28]. Available from: http://uwspace.uwaterloo.ca/handle/10012/5383

18. Ministry of Transportation Ontario (MTO). GreenPave: Ontario's First Pavement Sustainability Rating System. Road Talk: Ontario's Transportation Technology Transfer Digest [Internet]. 2010 [cited 2011 Mar 28];16(1). Available from: http://www.mto.gov.on.ca/english/transtek/roadtalk/rt1 6-1/#a6

19. Horvath A. PaLATE - Pavement Life-cycle Assessment Tool for Environmental and Economic Effects [Internet]. Consotrium on Green Design and Manufacturing: University of California, Berkeley. 2007 Jun 5 [cited 2011 Mar 13];Available from: http://www.ce.berkeley.edu/~horvath/palate.html

20. FHWA. FHWA Sustainable Highways Self-Evaluation Scorecard [Internet]. 2010 Oct [cited 2011 Mar 28];Available from: https://www.sustainablehighways.org/FHWA_Sustaina ble_Highways_Self-Evaluation_Tool_October_2010.pdf

21. Sustainable Sites Initiative. The Sustainable Sites Initiative [Internet]. [cited 2011 Mar 29];Available from: http://www.sustainablesites.org/

22. American Society of Landscape Architects, University of Texas at Austin, , and , United Stated Botanic Garden. The Sustainable Sites Initiative: Guidelines and Performance Benchmarks 2009 [Internet]. Sustainable Sites Initiative; 2009 [cited 2011 Mar 29]. Available from: http://www.sustainablesites.org/report/Guidelines%20a nd%20Performance%20Benchmarks_2009.pdf

23. Lee JC, Tuncer EB, Benson CH, Tinjum JM. Evaluation of Variables Affecting Sustainable Highway Design Using BE2ST-in-Highways System [Internet]. In: Proceedings of the 90th Annual Meeting of the Transportation Research Board. Washington, D.C.: Transportation Research Board (TRB); 2011 [cited 2011 Mar 29]. Available from: http://www.rmrc.unh.edu/Outreach/docs/Lee,Edil,Benson,Tinjum.pdf

24. Tuncer EB. Building Environmentally and Economically Sustainable Transportation Infrastructure-Highways (BE2ST-In-Highways) [Internet]. 2010 [cited 2011 Mar 29];Available from: http://itre.ncsu.edu/cte/EEConference/sessions/documents/08-3_Edil.pdf

25. Schendler A, Udall R. LEED is Broken...Let's Fix It [Internet]. 2005 [cited 2011 Mar 29];Available from: http://www.aspensnowmass.com/environment/images/LEEDisBroken.pdf

26. Lee J, Edil TB, Benson CH, Tinjum JM. Use if BEST In-Highways for Green Highway Construction Rating in Wisconsin [Internet]. In: Proceedings of The 2010 Green Streets And Highways Conference. Reston, VA: American Society of Civil Engineers (ASCE); 2010 [cited 2011 Mar 29]. p. 480-494.Available from: http://cedb.asce.org/cgi/WWWdisplay.cgi?271765

27. Saaty TL. The analytic hierarchy process: planning, priority setting, resource allocation. McGraw-Hill International Book Co.; 1980. 310 p.

28. Clark M, Paulli C, Tetreault Z, Thomas J. Green Guide for Roads Rating System [Internet]. Worcester, MA:

Worcester Polytechnical Institute (WPI); 2009 [cited 2011 Mar 30]. Available from: http://www.wpi.edu/Images/CMS/CEE/Green_Guide_f or_Roads_Rating_System.pdf

29. Greenalberta. Green Guide for Roads [Internet]. 2010 [cited 2011 Mar 29];Available from: http://greenalberta.ca/roads/

30. TAC. Green Guide for Roads Roadmap [Internet]. 2009 Jan 29 [cited 2011 Mar 30];Available from: http://www.tac-atc.ca/private/urban/pdfs/attachment4-0409.pdf

31. TAC. Green Guide for Roads - Project in Progress [Internet]. Transportation Association of Canada (TAC). 2011 Mar 30 [cited 2011 Mar 31];Available from: http://www.tac-atc.ca/english/projects/greenguide.cfm

32. Fung R. Transportation Association of Canada Urban Transportation Council: Green Guide for Roads - In Development [Internet]. 2010 Sep 26 [cited 2011 Mar 30];Available from: http://www.tac-atc.ca/private/urban/pdfs/att3-0910.pdf

33. TAC. Green Guide for Roads - Working Template [Internet]. 2010 Sep [cited 2011 Mar 30];Available from: http://www.tac-atc.ca/english/projects/pdf/greenguide.pdf

34. ASCE. New Organization to Develop Tool for Measuring Infrastructure Sustainability [Internet]. American Society of Civil Engineers (ASCE) Press Release. 2011 Feb 1 [cited 2011 Mar 31];Available from: http://www.asce.org/PressRelease.aspx?id=12884904365

35. Jones J. ILC Works to Help Develop Sustainability Rating System. ASCE Civil Engineering [Internet]. 2010 May [cited 2011 Mar 31];35(5). Available from: http://www.asce.org/PPLContent.aspx?id=2147488167

36. Shea D, Beard J. ISI: Institute for Sustainable Infrastructure [Internet]. 2011 Feb 23 [cited 2011 Mar 30];Available from: http://www.acec.org/advocacy/committees/pdf/eec0211_isi.pdf

37. Bowers CW, Beard J. Sustainable InfrastructureProject Rating System [Internet]. 2010 Jun 8 [cited 2011 Mar 30];Available from: http://www.acecmw.org/downloads/2010%20Annual%20Conference/Sustainable_Infrastructure_Project_Rating_System.pdf

38. IDOT, IRTBA, , and , ACEC-IL. I-LAST™: Illinois - Livable and Sustainable Transportation Rating System and Guide Version 1.01 [Internet]. Illinois DOT and Ilinois Joint Sustainability Group; 2010 [cited 2011 Mar 30]. Available from: http://www.dot.state.il.us/green/documents/I-LASTGuidebook.pdf

39. Hassan MM. SDFlex: A Framework for the Assessment and Construction of Sustainable Flexible Pavements. Journal of Green Building [Internet]. 2008 [cited 2011 Feb 4];3(3):108-118. Available from: http://www.atypon-link.com.proxy.lib.iastate.edu:2048/CPUB/doi/abs/10.3992/jgb.3.3.108

40. Oswald MR, McNeil S. Rating Sustainability: Transportation Investments in Urban Corridors as a Case Study. J. Urban Plng. and Devel. [Internet]. 2010

[cited 2011 Mar 31];136(3):177. Available from: http://ascelibrary.org/upo/resource/1/jupddm/v136/i3/p 177_s1?isAuthorized=no

41. Li L. Innovative Green Highway Construction Rating System in Mississippi [Internet]. Jackson State University. 2011 [cited 2011 Mar 27];Available from: http://imtransjsu.com/?page_id=596

Chapter 12:
Emerging Sustainable Pavement Materials and Technologies

Cementitious Materials

The production of portland cement is an extremely energy and resource intensive process and is a major contributor to GHG emissions. For instance, more than 1.5 tons of non-renewable raw materials are needed for producing a ton of portland cement and the production process releases about a ton of CO_2 in the environment. Thus, portland cement manufacturing has the highest carbon footprint and energy consumption in the production of concrete (1). Although, by increasing the thermal substitution rate of waste, the share of bioegenic wastes and the energy efficiency of cement kiln systems, the ecological impacts of cement production could be reduced, these involve cost-intensive changes in the production infrastructure (2).

During the manufacture of cement, CO_2 is produced from both the calcination of limestone and the combustion of fossil fuels to generate the high cement kiln temperatures needed to produce clinker. Clinker is the basic ingredient in the production of portland cement. Therefore, existing and emerging strategies for minimizing the overall carbon footprint associated with concrete pavements are focusing on reducing the amount of cement in a concrete mixture or reducing the amount of clinker needed to produce cement (3). Studies have shown that the use of cement types with a lower percentage of clinker has the potential to reduce the GHG emissions of concrete pavement up to 25% (2).

The amount of portland cement used in concrete can be reduced through one of the following existing approaches (4):

✓ Use of chemical admixtures

✓ Through mixture proportioning techniques

✓ Blending SCMs with portland cement

✓ Application of performance cements (ASTM C1157) that focus on performance and not on cement composition

Blended Hydraulic Cements

There is considerable interest in the uses of blended hydraulic cements and performance hydraulic cements owing to their inherent sustainability benefits (4). The term "hydraulic" cement is used to refer to cements (not just portland cements) that undergo chemical reactions with water and can harden even underwater or when constantly exposed to wet weather.

Blended hydraulic cements are specified under ASTM C595 and are produced by inter-grinding or blending portland cement with fly ash or natural pozzolans or slag cement, or creating a ternary combination of SCMs. The cement content of the blended cement can be reduced by as much as 50 percent through use of these SCMs. Especially, the use of very high volumes of fly ash in concrete (50 percent or more), resulting in High-Volume Fly Ash (HVFA) Concrete, is receiving renewed attention in the US owing to its sustainability benefits (5,6).

Performance-Specified Hydraulic Cements

The use of performance-specified hydraulic cements, with the recent adoption of ASTM C1157, encourages innovative development of composite portland cements (portland cement blended with multiple SCMs and/or limestone). They are based on the concept that direct material performance is of interest and not its composition (7). Additionally, non-portland-cement hydraulic binders that have the potential to significantly reduce the CO_2 associated with concrete production can be used under ASTM C1157 (4).

Portland-Limestone Cement (PLC)

Portland-Limestone Cement (PLC) is another modified hydraulic cement type that is manufactured by inter-grinding portland cement clinker with limestone at significantly higher limestone percentages than what is allowed by existing specifications. The current ASTM C150 specification allows up to 5 percent addition of limestone during the manufacture of portland cement. But, PLC, which allows up to 20 percent limestone, has been used in Europe for over 25 years and has been recently approved for use in Canada (8).

Next-Generation Cements

Alternative non-portland hydraulic cements that are being evaluated in this regard include rapid-setting, alkali-activated cements made from Class C fly ash and geopolymer cements made from kaolinite clays and Class F fly ash (4,7).

Another research area is looking into the development of cements that actually sequester CO_2 from atmosphere. For instance, there is considerable interest in a proprietary carbon-sequestering cement that sequesters CO_2 generated by power plants (flue gas) to form calcium and magnesium carbonates, which in turn may be used as aggregates or possibly as an SCM or cement (9).

Similarly, researchers are developing CO_2-negative cements and concretes that incorporate CO_2 in their structure (7,10). Proprietary eco-friendly cements (under the brand name eco-cement) which incorporate solid waste and sewage sludge, produced at lower kiln temperatures, which absorbs and sequesters CO_2 and at the same possessing rapid-hardening abilities are being investigated for highway applications (7). High-volume pozzolan cements (under the brand name Energetically Modified Cement) are also being considered among the candidate sustainable cement alternatives (11).

Concrete Materials

High-performance candidate concrete materials that are being examined for their potential sustainability advantages apart from their inherent beneficial properties for highway applications include Engineered Cement Composites (ECCs) (12), TiO_2-modified concrete (13), pervious concrete (14), Self-Consolidating or Self-Compacting Concrete (SCC) (15), precast concrete panels (16), sulfur concrete (17), Autoclaved Aerated Concrete (AAC) or Autoclaved Cellular Concrete (ACC) (18), geopolymer concrete (19), hydrophobic concrete (20), and ductile concrete (21).

Engineered Cement Composites (ECCs)

ECCs (also nicknamed "bendable" concrete) are micromechanically designed, highly ductile, fiber-reinforced cement-based materials that offer the potential for autogenous healing of hairline cracks (and are therefore known as self-healing concrete, see Fig. 12.1) apart from exhibiting improved structural integrity, resistance to plastic shrinkage, and improved post-cracking behavior (12,22). Some studies have reported that, compared to conventional concrete and HMA overlay systems, ECC overlay system reduces GHG emissions by 32% and 37%, life cycle energy consumption by 15% and 72%, and costs by 40% and 47%, respectively (23).

Self-Consolidating Concrete (SCC)

SCC (see Fig. 12.2) is a highly flowable, non-segregating, high-performance concrete that can effortlessly fill and consolidate in tight and constricted spaces and around congested steel rebars, without requiring mechanical vibration (15,25,26).

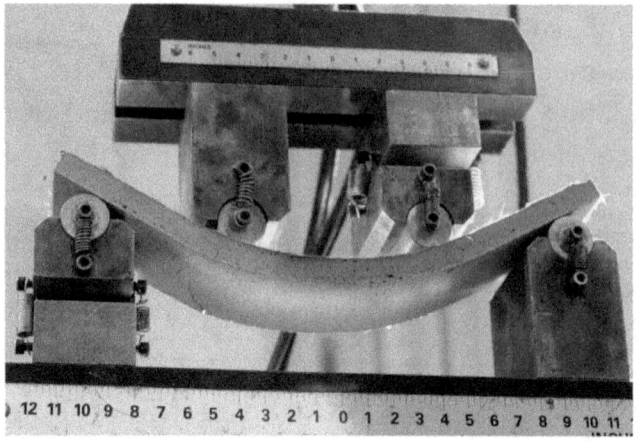

Fig. 12.1. Self-healing concrete (Engineered Cement Composites [ECC]) specimen bending as a force of 5% tensile strain is applied [*Source:* N. C. Moore, U-M News Service and College of Engineering (24)]

Geopolymer Concrete

Geopolymer represents a family of materials characterized by chains or networks of inorganic molecules with an amorphous microstructure (27). One class of geopolymers are comprised of aluminosilicate materials which has the potential to completely replace cement in concrete (27). These geopolymers are synthesized simply by mixing aluminosilicate reactive materials (which are either thermally activated natural materials like kaolinite clay or industrial byproducts such as fly ash or slag) and strongly alkaline solutions which polymerizes into molecular chains and networks to create the hardened binder.

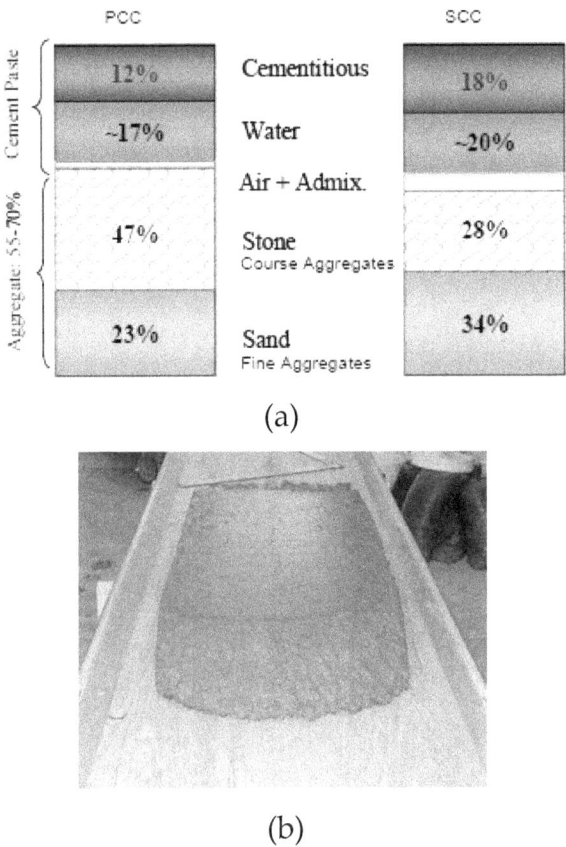

(a)

(b)

Fig. 12.2. (a) Relative proportions of mix components in PCC versus SCC [*Source:* FHWA (15)]; (b) Mini-Paver Test Section for SCC for Slip-Form Paving [*Source:* CP Tech Center (26)]

It has been shown that geopolymer cements (mentioned previously under Advanced Cementitious Materials), which come under the class of alkali-activated cements or inorganic polymer cements, has good volume stability than portland cement (28). Geopolymer concrete can achieve reasonable strength gain within a short time, has excellent durability, high fire resistance and low thermal conductivity (19).

It appears that geopolymer concrete technology is rapidly advancing in Europe and Australia although there are no widespread applications of it yet in transportation infrastructure (19). Geopolymer coated pavements are being used by U.S. military to resist the heat generated by vertical takeoff and landing aircraft (19).

Owing to the current limitations of geopolymer systems which include need for great care in their production, high alkalinity of the activating solution resulting in greater safety risk, increased energy consumption and GHG generation (because high alkalinity requires more processing), etc., its practical use in precast applications for transportation infrastructure is currently limited. Current research studies focus on addressing these issues and aim to create a low embodied energy, low CO_2 sustainable binder with geopolymers that has similar properties to portland cement (7,19).

Asphalt Paving Materials

Some of the promising materials investigated under this category include sulfur-extended asphalt (29), bio-derived asphalt binders, polymer-modified asphalt binders (30), Recycled Asphalt Shingles (RAS), etc.

Recycled Asphalt Shingles (RAS)

Currently, there is no requirement in any of the green rating systems (including the FHWA's rating system) to have a transportation project review prior to construction, although designers and contractors may choose to do so. In these systems, typically more points are awarded for sustainability strategies that will have greater positive impacts on energy efficiency and CO_2 emissions. An important point to keep in mind is that these green rating systems are constantly under development by design. Although the information presented here reflects the latest status with respect to each of them and are subject to revision, it is expected that the overall framework may not change significantly in a short time.

RAS are waste shingles (see Fig. 12.3) from old roofing applications salvaged for use in pavement construction materials (either as an aggregate supplement or in the modification of cold, warm and hot asphalt mixes) as well as in other construction applications (7). Asphalt Shingles typically consist of fiberglass or cellulose backing (2-15%); asphalt binder (19-22% on a fiberglass-matt base, 30-36% on a cellulose-felt base made with paper); sand-sized, ceramic-coated rock granules (20-38%); and mineral filler or stabilizer that includes limestone, dolomite and silica (8-40%) (31).

Fig. 12.3. Waste asphalt shingles [*Source:* WHRP (32)]

Waste shingles are obtained from two sources: (1) "tear off" or "post consumer" shingles obtained from the demolition of existing roofs; (2) "factory scrap" or "post industrial" shingles from factory rejects and tab cut-outs resulting from shingle production (7). Of the two, the latter provides better material consistency due to the uniformity of the source of shingle, although majority of the asphalt shingles disposed off annually in the US is from installation scraps and tear-offs from re-roofing (31). The use of RAS in CMA, WMA and HMA has significant benefits in terms of conservation of asphalt and aggregate materials, overall reduction in solid waste, reductions in energy requirements and GHG emissions associated with the production and placement of the mix (33).

Bio-based Asphalt

Renewable vegetable oil-based modifiers (formulations from soybean, corn, sunflower, and canola) have been investigated as possible modifiers for asphalt binders (7). Recent research studies are focusing on asphalt modifiers and extenders derived from co-products of lignocellulosic ethanol production (34) and development of bio-binders from bio-oils (oak flour, switchgrass, and corn stover) to be utilized as a direct alternative (100% replacement) to petroleum-based asphalt (referred to as bio-based asphalt) (35). Other research studies are studying bio-based asphalts derived from waste cooking oil (36), thermochemical conversion of swine manure (37), and urban yard waste carbonization (38).

Other Materials

Steel is used in most concrete pavements either in the form of reinforcement in the slabs, load transfer across joints using dowel bars, or longitudinal tie bars across adjacent pavement lanes. Candidate metallic and polymer materials investigated for their sustainability benefits in this regard include vitreous ceramic coatings for reinforcing steel, fiber-reinforced polymer bars for CRCPs, fiber-reinforced polymer dowel bars, zinc-clad dowel bars, and micro-composite steel for dowels and tie bars (7).

Advanced aggregate materials considered for use in asphalt concrete or PCC include synthetic aggregates manufactured using industrial waste material or by-products, manufactured aggregates using captured CO_2, and materials that allow internal concrete curing. The potential sustainability benefits of following materials are also being investigated: ultra-thin bonded wearing course or ultra-thin friction course for preventive maintenance, a proprietary lithium-based concrete curing material especially for hot, windy, low-humidity environments, a commercial soybean oil emulsion for concrete curing, workability-retaining concrete admixture, and concrete surface sealers which slow down the penetration of deicing chemicals (7).

Two-Lift Concrete Paving

Two-lift concrete paving allows the use of lower quality (lower durability or strength), locally available aggregate or recycled aggregate (such as RAP, concrete rubble, or local aggregate) in the bottom thick layer while the top thin layer is made up of high-quality aggregate designed to provide better resistance to freeze-thaw damage, reduced noise, or improved friction. Thus, instead of the homogeneous single layer commonly placed in concrete paving, two wet-on-wet layers or bonding wet to dry layers of concrete are placed during two-lift construction (39) (see Fig. 12.4). The FHWA is encouraging the use of two-lift concrete paving through its Pavement Technology and Highways for LIFE programs (40).

Although two-lift concrete paving technology has been around for more than a century, it is gaining increasing attention in the US owing to its sustainability benefits as well as advances made in materials knowledge and construction equipment, and increasing demands for pavement surfaces that meet specific noise, durability and safety objectives (41). It has continued to be a popular technique in Europe (especially in Austria, France, and Germany) to construct long-life pavements that use economical materials and have improved safety and pavement surface characteristics (42).

The reported benefits of two-lift paving process include (41):

✓ Reduced cost of materials and materials transportation
✓ Reduced environmental impact caused by quarrying and importing aggregates
✓ Increased societal benefits by lowering noise levels
✓ Increased opportunities for creating exposed aggregate surfaces (for quiet, safe, and long-lasting pavements) without increasing the cost or the environmental impact

✓ Long-life pavement that does not need frequent reconstruction

Since two-lift paving often requires the use of two plants, two slipform machines, and a special haul road, typically, the construction costs for two-lift pavements are about two times those of concrete pavements constructed using a standard one-life technique. However, recent advances in two-lift paving equipment and techniques combined with the opportunities that exist to use less expensive, local/recycled aggregates in the lower lift, the savings might be enough to offset part or all of the two-lift paving additional costs (39).

Photocatalytic Road Pavements

Although the primary pollution from vehicular traffic (i.e., fossil fuel combustion) is CO_2, other gasses are formed, such as NOx and SOx, that can lead to acid rain, smog, and respiratory issues (43). These high concentrations of nitrogen oxides and volatile organic compounds (VOCs) have been linked with serious public health hazards. According to the US EPA, 34% of the national NOx emissions come from road traffic (44). Nitrogen oxides can also travel long distances, causing a variety of health and environmental problems in locations far from their emissions source contributing to ozone and smog (44).

Titanium dioxide (TiO_2) has been used as a photochemical reactant in relation to environmental contaminants since the 1970s (45). It has been shown that TiO_2 can treat both organic pollutants and oxides such as NO, NO_2 and SO_2 at low concentration levels under UV radiation and has beneficial properties of air purification, self-cleaning, and photocatalytic anti-bacterial effect (46,47). During these reactions which occur at the surface of the built environment, the active ingredients (including titanium dioxide) are not consumed, but they act solely as catalysts. Since the early 1990s, TiO_2-based photocatalytic construction and building materials have been used widely used owing to their compatibility with traditional construction materials such as cement as well as their versatility to serve as both photocatalytic materials and structural materials (47).

Since pavements have an abundant surface area and are proximal to a major source of air pollution (from road traffic), TiO_2-based photocatalytic pavements are being researched for their environmental effectiveness. The use of TiO_2-blended concrete materials for pavement applications began in 1999 with the development of an interlocking paving block with an oxidative NO_x removal capability (48). Since then several research studies have been carried out to investigate the air purifying efficiency and durability of photocatalytic concrete pavements (49-53).

A variety of proprietary TiO_2-based concrete products and application mechanisms are already available on the European market and their working mechanism is proven under laboratory conditions (52). A commonly used method of integrating the TiO_2 particles within the pavement surface consists of blending these particles with a cementitious mortar mixture and then applying it as an ultra-thin coating on the concrete pavement surface (51).

Currently, the following research topics are being actively pursued in this area:

✓ Environmental effectiveness of photocatalytic asphalt pavement surfaces incorporating ultrafine/nano titanium dioxide (54,55)

✓ Development of photocatalytic Warm-Mix Asphalt (WMA) mixtures (56)

✓ Effectiveness of photocatalytic concrete for use in two-lift pavements, barrier walls, and sound walls (43)

Energy-Harvesting Pavements

Researchers are currently looking into different ways of harvesting the energy absorbed by pavements to melt ice on roads during winter, power streetlights, illuminate signs, heat buildings and potentially use it for many other purposes:

✓ Harvesting kinetic energy from paving slabs in high footfall environments and converting into electricity (57)

✓ Harvesting solar energy from asphalt pavements and reducing the UHI (58,59)

✓ Harvesting solar energy from asphalt pavements through one of the approaches: (a) wrapping flexible photovoltaic cells around the top of Jersey barriers dividing highways to provide electricity; (b) embedding water filled pipes beneath the asphalt; and (c) embedding thermo-electric materials in the roadway at different depths (60)

References

1. Marceau M, Nisbet M, VanGeem M. Life Cycle Inventory of Portland Cement Concrete [Internet]. Skokie, IL: Portland Cement Association (PCA); 2007 [cited 2011 Feb 25]. Available from: http://assets.ctlgroup.com/a4c9c83b-c381-4fda-baab-16cf64bbcde5.PDF

2. Gschosser F, Bosch M, Wallbaum H. Hidden Ecological Potentials in Concrete Pavements [Internet]. In: Proceedings of the CIB World Congress. Salford Quays, UK: 2010 [cited 2011 Mar 17]. Available from: http://www.ibi.ethz.ch/nb/publications/papers/201005_Gschoesser.et.al_Hidden.ecological.potentials.pdf

3. Grove J, Vanikar S, Wathne L. Sustainability Opportunities for Concrete Pavements [Internet]. In: Proceedings of the International Conference on Sustainable Concrete Pavements: Practices, Challenges, and Directions. Sacramento, California: Federal Highway Administration (FHWA); 2010 [cited 2011 Feb 25]. p. 55-69.Available from:

http://www.fhwa.dot.gov/pavement/concrete/2010acptc onf.pdf

4. FHWA. Geopolymer Concrete [Internet]. Washington, D.C.: Federal Highway Administration (FHWA); 2010 [cited 2011 Feb 25]. Available from: http://www.fhwa.dot.gov/pavement/concrete/pubs/hif1 0014/hif10014.pdf

5. Malhotra VM, Mehta PK. High-Performance, High-Volume Fly Ash Concrete for Building Sustainable & Durable Structures. Third Edition. Ottawa, Canada: Supplementary Cementing Materials for Sustainable Development Inc.; 2008. 120 p.

6. Mehta PK. High-Performance, High-Volume Fly Ash Concrete for Sustainable Development [Internet]. In: Proceedings of the International Workshop on Sustainable Development and Concrete Technology. Beijing, China: Iowa State University; 2004 [cited 2011 Feb 25]. p. 3-14.Available from: http://www.intrans.iastate.edu/pubs/sustainable/mehta sustainable.pdf

7. Tayabji S, Smith KD, Van Dam T. Advanced High-Performance Materials for Highway Applications: AReport on the State of Technology [Internet]. Washington, D.C.: Federal Highway Administration (FHWA); 2010 [cited 2011 Feb 25]. Available from: http://www.fhwa.dot.gov/pavement/materials/pubs/hif 10002/hif10002.pdf

8. CAC. Backgrounder: Portland-Limestone Cement [Internet]. Ontario, Canada: Cement Association of Canada (CAC); 2009 [cited 2011 Feb 25]. Available from:

http://www.cement.ca/images/stories/plc_bkgndr-_eng-
_sept._1_2009.pdf

9. Clodic L, Patterson J, Ryan C, Holland T. Next
    Generation Paving Materials UsingMineralized CO2
    Captured from Flue Gas [Internet]. In: Proceedings of
    the International Conference on Sustainable Concrete
    Pavements: Practices, Challenges, and Directions.
    Sacramento, California: Federal Highway
    Administration (FHWA); 2010 [cited 2011 Feb 25]. p.
    359-370.Available from:
    http://www.fhwa.dot.gov/pavement/concrete/2010acptc
    onf.pdf

10. Gaurdian News and Media Limited. Revealed: The
    cement that eats carbon dioxide [Internet]. 2008 Dec 31
    [cited 2011 Feb 26];Available from:
    http://www.guardian.co.uk/environment/2008/dec/31/c
    ement-carbon-emissions

11. Ronin V. An Industrially Proven Solution for
    Sustainable Pavements of High-VolumePozzolan
    Concrete—Using Eneretically Modified Cement, EMC
    [Internet]. In: Proceedings of the International
    Conference on Sustainable Concrete Pavements:
    Practices, Challenges, and Directions. Sacramento,
    California: Federal Highway Administration (FHWA);
    2010 [cited 2011 Feb 25]. Available from:
    http://www.fhwa.dot.gov/pavement/concrete/2010acptc
    onf.pdf

12. Li VC. On Engineered Cementitious Composites (ECC):
    A Review of the Material and its Applications. Journal
    of Advanced Concrete Technology [Internet].

2003;1(3):215–230. Available from: http://www.j-act.org/4-3.html

13. Jayapalan AR, Lee BY, Fredrich SM, Kurtis KE. Influence of Additions of Anatase TiO2 Nanoparticles on Early-Age Properties of Cement-Based Materials. Transportation Research Record: Journal of the Transportation Research Board [Internet]. 2010 [cited 2011 Feb 26];2141:41-46. Available from: http://pubsindex.trb.org/view.aspx?id=917947

14. NRMCA. Pervious Concrete Pavement for Green, Sustainable Porous and Permeable Stormwater Drainage [Internet]. National Ready Mixed Concrete Association (NRMCA). 2010 [cited 2011 Feb 26];Available from: http://www.perviouspavement.org/

15. FHWA. Self-Consolidating Concrete [Internet]. Washington, D.C.: Federal Highway Administration (FHWA); 2008 [cited 2011 Feb 26]. Available from: http://www.fhwa.dot.gov/hfl/innovations/pdfs/scc.pdf

16. FHWA. Precast Concrete Pavement Systems [Internet]. Washington, D.C.: Federal Highway Administration (FHWA); 2010 [cited 2011 Feb 27]. Available from: http://www.fhwa.dot.gov/hfl/innovations/pdfs/precast.pdf

17. ACI. Guide for Mixing and Placing Sulfur Concrete inConstruction. [Internet]. Farmington Hills, MI: American Concrete Institute (ACI); 1993 [cited 2011 Feb 26]. Available from: http://www.concrete.org/PUBS/JOURNALS/OLJDetails.asp?Home=MJ&ID=2150

18. Portland Cement Association (PCA). Autoclaved Cellular Concrete – theBuilding Material of the 21st

Century. Concrete Technology Today [Internet]. 1991 Jul [cited 2011 Feb 26];12(2):1-3. Available from: http://www.cement.org/tech/pdfs/PL912PL921AAC.pdf

19. FHWA. Geopolymer Concrete [Internet]. Washington, D.C.: Federal Highway Administration (FHWA); 2010 [cited 2011 Feb 26]. Available from: http://www.fhwa.dot.gov/pavement/concrete/pubs/hif1 0014/hif10014.pdf

20. CP J. Hydrophobic Concrete Sheds Waterproofing Membrane. Concrete Products (CP) [Internet]. 2006 Jan [cited 2011 Feb 26];109(1). Available from: http://www.hycrete.com/files/media-releases/Hycrete-Media-01-01-06.pdf

21. CIF. Ultra-High Performance Ductile Concrete [Internet]. 2007 [cited 2011 Feb 26];Available from: http://www.cif.org/noms/2007/09_-_Ultra-High_Performance_Ductile_Concrete.pdf

22. Li VC. Engineered Cementitious Composites (ECC) – Material, Structural, and Durability Performance. In: Nawy E, editor. Concrete Construction Engineering Handbook. Boca Raton, Florida: CRC Press; 2008. p. 40.

23. Zhang H. Sustainable Pavement Asset Management Based on Life CycleModels and Optimization Methods [Internet] [Ph.D. Dissertation]. [Ann Arbor, MI]: University of Michigan; 2009 [cited 2011 Mar 16]. 147 p.Available from: http://css.snre.umich.edu/css_doc/CSS09-14.pdf

24. Moore NC. Bendable self-healing concrete [Internet]. University of Michigan, Ann Arbor. 2009 [cited 2011 Feb 26];Available from:

http://www.engin.umich.edu/newscenter/feature/selfhe
alconcrete/bendable_concrete_0/index.html

25. NRMCA. Self Consolidating Concrete [Internet].
    National Ready Mixed Concrete Association (NRMCA).
    2007 [cited 2011 Feb 26];Available from:
    http://www.selfconsolidatingconcrete.org/Advanced%2
    0Topics/advanced%20topics.htm

26. Wang K, Shah SP, White DJ, Gray J, Voigt T, Gang L, et
    al. Self-Consolidating Concrete—Applications for Slip-
    Form Paving: Phase I (Feasibility Study) [Internet].
    Ames, IA: National Concrete Pavement Technology
    Center, Iowa State University; 2005 [cited 2011 Feb 26].
    Available from:
    http://www.intrans.iastate.edu/reports/scc_pave.pdf

27. Davidovits J. Geopolymer: Chemistry & Applications
    [Internet]. Second Edition. Saint-Quentin, France:
    Geopolymer Institute; 2008 [cited 2011 Feb 26]. 594
    p.Available from:
    http://www.geopolymer.org/learning/book-
    geopolymer-chemistry-and-applications

28. Rangan BV. Low-Calcium, Fly-Ash-Based Geopolymer
    Concrete. In: Nawy E, editor. Concrete Construction
    Engineering Handbook. Boca Raton, Florida: CRC
    Press; 2008.

29. Timm D, Tran N, Taylor A, Robbins M, Powell B.
    Evaluation of Mixture Performance and Structural
    Capacity of Pavements Using Shell Thiopave: Phase I -
    Mix Design, Laboratory Performance Evaluation and
    Structural Pavement Analysis and Design [Internet].
    Auburn, AL: National Center for Asphalt Technology

(NCAT); 2009 [cited 2011 Feb 26]. Available from: http://www.ncat.us/reports/rep09-05.pdf

30. Bahia HU, Hanson DI, Zeng M, Zhai H, Khatri MA, R.M. Characterization of Modified Asphalt Binders in Superpave Mix Design [Internet]. Washington, D.C.: Transportaion Research Board (TRB); 2001 [cited 2011 Feb 26]. Available from: http://onlinepubs.trb.org/onlinepubs/nchrp/nchrp_rpt_4 59-a.pdf

31. NERC. Asphalt Shingles Waste Management in the Northeast: Fact Sheet [Internet]. 2007 Feb [cited 2011 Feb 26];Available from: http://www.nerc.org/documents/asphalt.pdf

32. WHRP. WHRP E-Newsletter [Internet]. Wisconsin Highway Research Program (WHRP). 2010 Jun [cited 2011 Feb 27];Available from: http://www.whrpnews.org/06-10/06-10.htm

33. Robinette C, Epps J. Energy, Emissions, Material Conservation, and Prices Associated with Construction, Rehabilitation, and Material Alternatives for Flexible Pavement. Transportation Research Record: Journal of the Transportation Research Board [Internet]. 2010 Dec 1 [cited 2011 Feb 27];2179:10-22. Available from: http://dx.doi.org/10.3141/2179-02

34. McCready NS, Williams R. Utilization of Biofuel Coproducts as Performance Enhancers in Asphalt Binder. Transportation Research Record: Journal of the Transportation Research Board [Internet]. 2008 Dec 1 [cited 2011 Feb 26];2051:8-14. Available from: http://dx.doi.org/10.3141/2051-02

35. Raouf MA, Williams RC. Determination of Pre-Treatment Procedure Required for Developing Bio-Binders from Bio-Oils [Internet]. In: Proceedings of the 2009 Mid-Continent Transportation Research Symposium. Ames, IA: Iowa State University; 2009 [cited 2011 Feb 26]. p. 14.Available from: http://www.intrans.iastate.edu/pubs/midcon2009/Raouf Bio.pdf

36. WSU. Developing An Earth-Friendly Asphalt [Internet]. Washington State University (WSU). 2010 [cited 2011 Feb 26];Available from: http://www.cea.wsu.edu/environment/bioasphalt

37. Fini EH, Yang, S-H, Xiu S. Characterization and Application of Manure-Based Bio-binder in Asphalt Industry [Internet]. In: TRB 89th Annual Meeting Compendium of Papers DVD. Washington, D.C.: Transportation Research Board (TRB); 2010 [cited 2011 Feb 26]. p. 14.Available from: http://trid.trb.org/view.aspx?id=910772

38. TRB Research in Progress. Bioasphalt from Urban Yard Waste Carbonization [Internet]. Transportation Research Board (TRB). 2010 Mar [cited 2011 Feb 26];Available from: http://rip.trb.org/browse/dproject.asp?n=25572

39. Cable JK. ReassessingTwo-Lift Paving [Internet]. Ames, IA: National Concrete Pavement Technology Center, Iowa State University; 2004 [cited 2011 Feb 24]. Available from: http://publications.iowa.gov/2958/1/TwoLiftSummary.pdf

40. FHWA Highways for LIFE. States Take Another Look at Two-Lift Paving. INNOVATOR: Accelerating Innovation for American Driving Experience [Internet]. 2007 Dec [cited 2011 Feb 25];V3(4). Available from: http://www.fhwa.dot.gov/hfl/innovator/issue04.cfm

41. Shields-Cook S, Taylor P. Road Construction: Working a double lift. Roads & Bridges [Internet]. 2009 Jan [cited 2011 Feb 24];:32-35. Available from: http://www.roadsbridges.com/articles/workingadoublel ift.pdf

42. Hall K, Dawood D, Vanikar S, Tally R, Cackler ET, Correa A, et al. Long-Life Concrete Pavements in Europeand Canada [Internet]. Washington D.C.: Federal Highway Administration (FHWA); 2007 [cited 2010 Feb 6]. Available from: http://international.fhwa.dot.gov/pubs/pl07027/pl07027. pdf

43. PCA. Self-Cleaning Concrete [Internet]. Portland Cement Association (PCA). [cited 2011 Feb 25];Available from: http://www.cement.org/tech/self_cleaning.asp

44. US EPA. Nitrogen Oxides - Mobile Source Emissions [Internet]. US EPA Mobile Source Emissions - Past, Present, and Future. 2010 [cited 2011 Feb 25];Available from: http://www.epa.gov/otaq/invntory/overview/pollutants /nox.htm

45. Fujishima A, Honda K. Electrochemical Photolysis of Water at a Semiconductor Electrode. Nature [Internet]. 1972 Jul 7 [cited 2011 Feb 25];238(5358):37-38. Available from: http://dx.doi.org/10.1038/238037a0

46. Agrios A, Pichat P. State of the art and perspectives on materials and applications of photocatalysis over TiO2. Journal of Applied Electrochemistry [Internet]. 2005 Jul 17;35(7):655-663. Available from: http://dx.doi.org/10.1007/s10800-005-1627-6

47. Chen J, Poon C-sun. Photocatalytic construction and building materials: From fundamentals to applications. Building and Environment [Internet]. 2009 Sep [cited 2011 Feb 25];44(9):1899-1906. Available from: http://www.sciencedirect.com/science/article/B6V23-4VFK7T2-1/2/e2ea68542489ae23aca67b2e74e6ddd4

48. Murata Y, Tawara H, Obata H, Takeuchi K. Air Purifying Pavement: Development of Photocatalytic Concrete Blocks. Journal of Advanced Oxidation Technologies [Internet]. 1999 Jul;4(2):227-230. Available from: http://www.ingentaconnect.com/content/stn/jaots/1999/00000004/00000002/art00019

49. Ballari MM, Hunger M, Hüsken G, Brouwers HJH. NOx photocatalytic degradation employing concrete pavement containing titanium dioxide. Applied Catalysis B: Environmental [Internet]. 2010 Apr 6 [cited 2011 Feb 25];95(3-4):245-254. Available from: http://www.sciencedirect.com/science/article/B6TF6-4Y5GXWP-1/2/9f28da26dd6bc3cac809e49241415874

50. Dylla H, Hassan MM, Mohammad LN, Rupnow T, Wright E. Evaluation of Environmental Effectiveness of Titanium Dioxide Photocatalyst Coating for Concrete Pavement. Transportation Research Record: Journal of the Transportation Research Board. 2010;2164:46–51.

51. Hassan MM, Dylla H, Mohammad L, Rupnow T. Evaluation of the durability of titanium dioxide photocatalyst coating for concrete pavement. Construction and Building Materials [Internet]. 2010 Aug [cited 2011 Feb 25];24(8):1456-1461. Available from: http://www.sciencedirect.com/science/article/B6V2G-4YBVN2S-2/2/cfaa2c82a88a16648192e7fa6785e7e1

52. Hüsken G, Hunger M, Brouwers HJH. Experimental study of photocatalytic concrete products for air purification. Building and Environment [Internet]. 2009 Dec [cited 2011 Feb 25];44(12):2463-2474. Available from: http://www.sciencedirect.com/science/article/B6V23-4W6Y33N-1/2/aba90eb8db0cd901df781f85b770bb7f

53. Poon CS, Cheung E. NO removal efficiency of photocatalytic paving blocks prepared with recycled materials. Construction and Building Materials [Internet]. 2007 Aug [cited 2011 Feb 25];21(8):1746-1753. Available from: http://www.sciencedirect.com/science/article/B6V2G-4M0BHDH-3/2/26db229a62126e2902de06baf334ae36

54. TRB Research in Progress. Environmental Effectiveness of Photocatalytic Asphalt Pavement Surfaces Incorporating Ultrafine/Nano Titanium Dioxide [Internet]. Transportation Research Board (TRB). 2010 Jan [cited 2011 Feb 25];Available from: http://rip.trb.org/browse/dproject.asp?n=27918

55. Berthelot A. Concrete Steps: LSU Professor to Lay Pollution-Cleaning Pavement on Campus and Aster Street Dec. 13 [Internet]. LSu Media Center. 2010 Dec 10 [cited 2011 Feb 25];Available from:

http://www.lsu.edu/ur/ocur/lsunews/MediaCenter/Ne
ws/2010/12/item22709.html

56. Hassan MM, Mohammad L, Dylla H, Cooper S,
     Mokhtar A, Asadi S. A Breakthrough Concept in the
     Preparation of Highly-SustainablePhotocatalytic Warm
     Asphalt Mixtures [Internet]. In: Proceedings of 2011
     NSF Engineering Research and Innovation Conference.
     Atlanta, Georgia: National Science Foundation (NSF);
     2011 [cited 2011 Feb 25]. p. 6.Available from:
     http://www.pureti.com/pdfs/PURETI+on+Asphalt+-
     +NSF11-Paper+f+Nov+2010+LSU.pdf

57. Pavegen Systems Limited. Generating Energy from
     Your Footsteps [Internet]. 2009 [cited 2011 Feb
     27];Available from: http://www.pavegensystems.com/

58. Mallick RB, Chen B-L, Bhowmick S, Hulen MS.
     Capturing Solar Energy from Asphalt Pavements
     [Internet]. In: Proceedings of the 2nd International
     Society for Asphalt Pavements (ISAP) Symposium on
     Asphalt Pavements and Environment. Zurich,
     Switzerland: International Society for Asphalt
     Pavements (ISAP); 2008 [cited 2011 Feb 26]. Available
     from: http://users.wpi.edu/~rajib/Pavement%20Energy-
     ISAP-August%202008.pdf

59. Mallick RB, Chen BL, Bhowmick S. Harvesting energy
     from asphalt pavements and reducing the heat island
     effect. International Journal of Sustainable Engineering
     [Internet]. 2009 [cited 2011 Feb 26];2(3):214–228.
     Available from: http://users.wpi.edu/~rajib/Draft-
     2White-Paper-on-Reduce-Harvest-Heat-from-
     Pavements-Nov-2008.pdf

60. McLeish T. URI researchers aim to harvest solar energy from pavement to melt ice, power streetlights, heat buildings [Internet]. University of Rhode Island. 2010 Nov 9 [cited 2011 Feb 27];Available from: http://www.uri.edu/news/releases/?id=5584

www.ingramcontent.com/pod-product-compliance
Lightning Source LLC
Chambersburg PA
CBHW070224190526
45169CB00001B/67